PLATYPUS

AUSTRALIAN NATURAL HISTORY SERIES

PLATYPUS

FOURTH EDITION

TOM GRANT

ILLUSTRATED BY DOMINIC FANNING

CSIRO
PUBLISHING

National Library of Australia Cataloguing-in-Publication entry

Grant, Tom, 1947– .
Platypus.

4th ed.
Bibliography.
Includes index.
ISBN 9780643093706 (pbk.).

1. Platypus. I. Fanning, Dominic. II. Title.
(Series: Australian natural history series).

599.29

Available from
CSIRO PUBLISHING
150 Oxford Street (PO Box 1139)
Collingwood VIC 3066
Australia

Telephone: +61 3 9662 7666
Local call: 1300 788 000 (Australia only)
Fax: +61 3 9662 7555
Email: publishing.sales@csiro.au
Web site: www.publish.csiro.au

Set in 10.5/14 Palatino
Cover and text design by James Kelly
Typeset by Barry Cooke Publishing Services
Printed in Australia by Ligare

CONTENTS

Preface		vii
1	Introduction	1
2	Breeding biology	13
3	Spurs and venom glands	41
4	The sensory world of the platypus	51
5	Energetics, diving and foraging	63
6	Ecology	81
7	Ancestry and evolution	99
8	Conservation: platypuses and people	107
9	Questions, answers and misconceptions	131
Selected references		145
Index		157

PREFACE

When I left the University of Canterbury in New Zealand at the end of the 1960s, my knowledge of the platypus was confined to its brief mention in vertebrate biology lectures and several diagrams, with associated text, in JZ Young's 1962 edition of *The Life of Vertebrates*. Little did I know then that, within two years of graduating in Zoology, I would begin to study this unique Australian animal and continue to do so for the next 35 years. The first UNSW Press edition of *The Platypus* was published in 1984. This hard-cover volume featured Dominic Fanning's excellent line drawings reproduced on art-quality paper. Revised editions of the book, entitled *The platypus: A unique mammal*, were published in 1989 and 1995 as part of the UNSW Press Natural History Series. In the previous volumes, there was an introductory chapter, followed by four additional chapters dealing with the biology of the platypus in response to seasonal changes. The current edition takes a more traditional approach, based on major subject areas. The final chapter is a series of frequently asked questions (FAQ), divided into subject areas/chapters, which give the reader quick access to particular areas of interest. Like the previous editions and many of the other books in the Natural History Series, the current edition is aimed predominantly at the layperson and references are not cited in the text. However, for readers who wish to find out more detail, a chapter-by-chapter bibliography is supplied at the end of the book. Using this bibliography, a reader should be able to locate detailed reference material, either by consulting the references themselves or by going to their bibliographies, a certain number of which are review articles or books.

There are numerous people without whom neither this nor the previous editions of *The Platypus* could have been written. The first to mention are, of course, the relatively small group of researchers and many volunteers whose dedicated work produced the subject material upon which the book is based. Several individuals deserve special mention. My wife, Gina, whose life and aspirations over the years have been frequently affected by the whims of *Ornithorhynchus anatinus*, deserves special thanks. She has supported me throughout the research, the field work and the writing. My very patient brother, Donald Grant, has read, commented on and improved all four volumes of the book. Peter and Meredith Temple-Smith also read and reviewed this and the previous editions. Peter has been a valued friend and colleague, who has had an ongoing interest in the platypus since the early 1970s. I also thank the following people for generously finding time to read and comment on various drafts of particular chapters or parts of chapters – Mike Archer, Philip Bethge, Sue Hand, Margaret Hawkins, Philip Kuchel, Sarah Munks, Keith Payne, Uwe Proske and Melody Serena.

In this and previous volumes, it is the illustrative and photographic material that greatly highlights the writing. Other than three specified items, Dominic Fanning produced all of the illustrative material, a good deal of which is new, but some of which includes the excellent line drawings that so beautifully embellished the original volume of the book. Where unspecified in the legend, the photographs are my own, but the following people or organisations are also thanked for generously donating the use of their photographs and/or other illustrative material: Australian Geographic/Peter Aitchison, Faye Bedford, Ros Bohringer, Joanne Connolly, Foster's Group Limited, Bob McBlain, John Matthews, NSW Department of Lands, Russell Millard, Uwe Proske, Matt Ryan, Ederic Slater, Jenny Taylor/Riversleigh Project, Peter Temple-Smith, Peter Tonelli and Richard Whittington. John Matthews also kindly provided great help in the copying and preparation of photographic material. Ian Montgomery – a dedicated ornithologist, professional bird photographer and long-time friend – only with great reluctance agreed to take photographs of a mammal. Several of his photographs have been used in the book but the stunning image of the platypus on the front cover eerily captures the special character of this animal in the wild.

My late friend and colleague, Merv Griffiths, to whom this edition of the book is dedicated, encouraged and inspired numerous platypus researchers during the many years he was involved in the study of monotreme biology. His musing on the banks of the upper Shoalhaven River one evening around 1989, 'perhaps this was the animal of all time', has become a phrase frequently used by those privileged enough to have worked with the platypus.

Tom Grant

1
INTRODUCTION

'.... the animal of all time'.

MERV GRIFFITHS (ca 1989)

When first discovered by Europeans, the platypus was thought to be a fake. Various dried or pickled specimens, reaching Britain and Europe at the end of the 19th century, were closely examined by naturalists and scientists determined to find the places where the different bits had been stitched together by a wily antipodean taxidermist. None were found and gradually, but with considerable consternation, acrimony and argument among eminent biologists of the time, the species was finally recognised as a unique mammal indeed, perhaps the 'animal of all time'.

The Monotremes: the egg-laying mammals

The Class Mammalia consists of two distinct sub-classes, those that give birth to live young (Sub-class Theria – the eutherian or placental and the marsupial mammals) and those that lay eggs (Sub-class Prototheria – the monotremes). There are two living families of monotremes: the echidnas or spiny ant-eaters (Family Tachyglossidae) and the platypus (Family

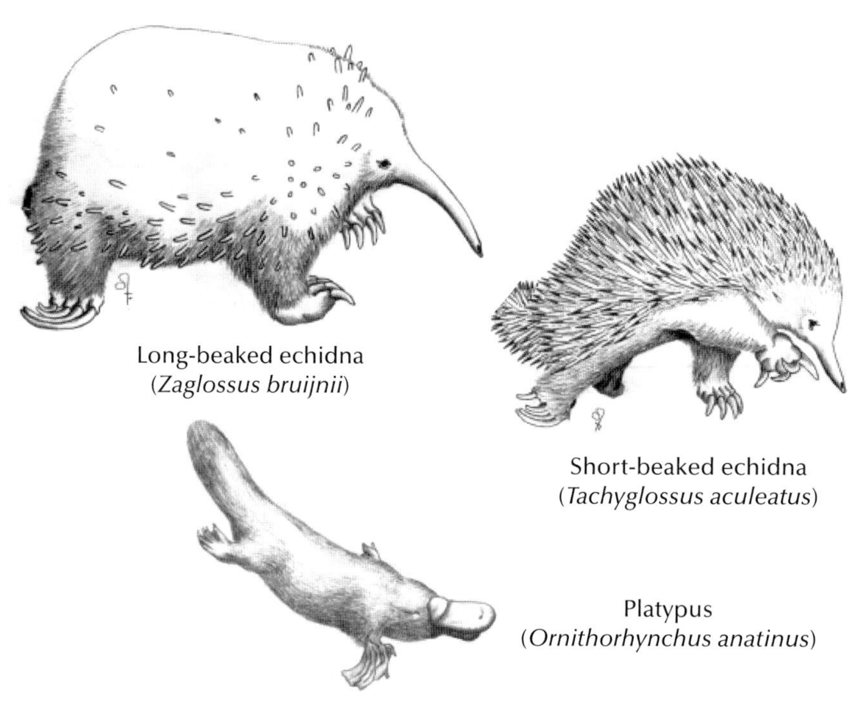

Figure 1.1 The three most common species of monotremes have similarities but the platypus is very different from the two species of echidna.

Ornithorhynchidae). Although the long-beaked and short-beaked echidnas and the platypus are the only three commonly known monotremes (Figure 1.1), additional rare species of echidna have been recently recognised to occur in New Guinea. But even including these less common species, the modern monotremes belong to a very small and unique group.

The long-beaked echidnas occur only in New Guinea, where they inhabit alpine meadows and humid montane forests. Their distribution and abundance are poorly understood and even the most well-known species, *Zaglossus bruijnii,* is considered by the International Union for Conservation of Nature and Natural Resources (IUCN) to be endangered, due to the clearing of its forest habitat for agriculture and as a result of hunting using dogs and modern firearms. Long-beaked echidnas, in fact, are neither ant-eaters nor particularly spiny. They eat mainly worms, centipedes, beetles and other terrestrial insects, and have only a few spines. The short-beaked echidna (*Tachyglossus aculeatus*), on the other hand, is very spiny and feeds predominantly on ants and termites, a ubiquitous food source that occurs throughout its distribution in Australia and parts of New Guinea.

Externally, the short-beaked and long-beaked echidnas appear quite different from each other, and both appear very different from the single living platypus species, *Ornithorhynchus anatinus*. However, they all share their egg-laying habit and have a good number of internal similarities, particularly their skeletons, which place them in the Order Monotremata. These unique species, previously thought to be a first evolutionary 'experiment' of the mammalian condition, are in fact highly adapted remnants of a once much more diverse radiation of egg-laying mammals.

Despite their unusual egg-laying reproduction, more often associated with birds or reptiles, the monotremes are definitely considered to be mammals; they suckle their young on milk and have fur- or hair-covered bodies (including spines in the echidnas which are modified hairs). The bones of the pectoral and pelvic girdles are quite similar to those of certain reptiles, but the rest of their skeletons are definitely mammalian. The lower jaw consists of a single pair of bones, the dentaries, which form a hinge with the squamosal bone in the skull, as occurs in all mammals. The inner ear is also mammalian, having three sound-conducting ossicles: the malleus, incus and stapes.

The platypus

Several other species of small vertebrates that occupy various waterways of eastern Australia can be mistaken for the platypus. Figure 1.2 shows that, when resting on the surface of the water between foraging dives, the platypus presents a very low profile, often making it very difficult to see. During the dive, the back is arched as the animal thrusts downwards and forwards. This dive creates a spreading ring, similar to, but subtly different from, the one left by a fish jumping. When the animal surfaces again to breathe, there is also a ring in the water, which frequently catches the light in the rising or setting sun in the morning or late afternoon. When this ring appears, however, the three small humps of head, back and tail remain visible, confirming the presence of a platypus. Occasionally the platypus will swim on the surface and create a low bow wave (see Colour Plate (top), page 33). The native water rat (*Hydromys chrysogaster*) also creates a bow wave when swimming, but this wave is more pronounced than that of the platypus. The water rat's body moves in a sinuous manner and it does not arch its back when diving. It also has a long tail, often with a visible white tip, and the external ears can normally be seen when the animal is swimming on the surface. The eastern water dragon (*Physignathus leseuerii*)

Figure 1.2 The profile of the platypus in the water often makes it difficult to be seen from the stream bank. Photo: © Ian Montgomery, birdway.com.au

overlaps much of the distribution of the platypus in eastern Queensland, New South Wales and into the north-east of Victoria. However, this species is more easily distinguished from the platypus in the water. The water dragon holds its head quite high in the water. There is no diving action, the animal simply seems to slip below the surface. Its swimming is noticeably much more sinuous in nature than either that of the water rat or the platypus. People sometimes report hearing a platypus 'plop' into the water. A platypus seldom 'plops' – it slides into the water. Those who hear such a noise are more likely to be reporting a freshwater turtle dropping into the water from a log or rock, where they often bask in the sun. Introduced carp (*Cyprinus carpio*) rolling at the surface, cormorants fishing and some other water birds diving are sometimes mistaken for a platypus. Most often these misidentifications are made by people who have not seen a platypus in the wild. Once seen, the profile and distinctive diving action of *Ornithorhynchus anatinus* are seldom forgotten.

Physical features

No other animal on Earth looks quite like the platypus. Figure 1.3 shows the most notable of its external features and Figure 1.4 highlights the unique nature of its internal skeleton.

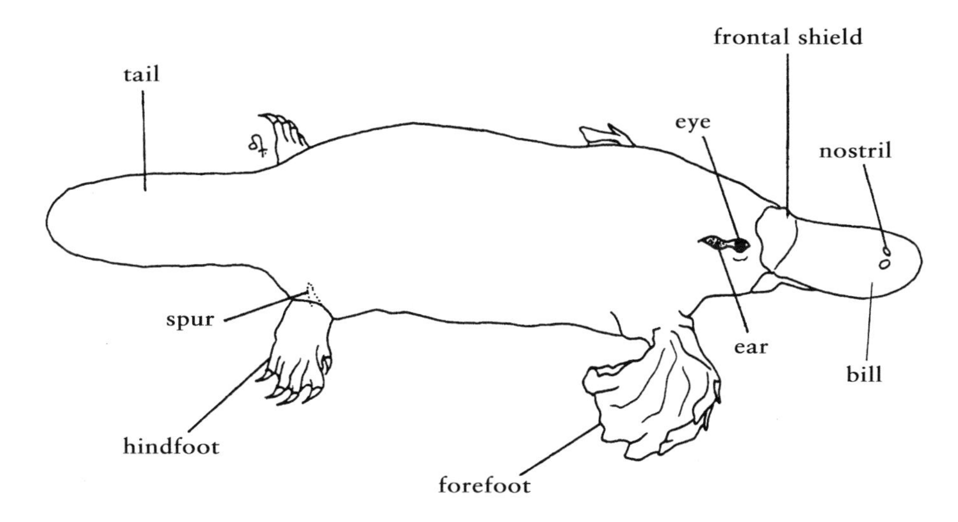

Figure 1.3 The main external features of the platypus.

Of course, the most distinctive feature is the bill, which is not hard like the bill of a duck, but soft and pliable. It is well supplied with nerves and is used by the animal to locate food and to find its way around under water. Behind the bill on either side of the head are two grooves that house the ear openings and the eyes which close when the platypus dives (see Colour Plate (bottom), page 33). The lower bill is smaller than the upper and is supported by the single pair of elongated dentary bones (Figure 1.4). The nostrils are on the top of the bill, not far back from the tip. They are raised slightly and the bill can be extended upwards so the animal is able to breathe at the surface when the rest of its body is almost completely submerged.

Platypuses are not as large as most people expect, with an adult female being much smaller than the average household cat (Figure 1.5). A very large male can be 60 cm from tip of bill to tip of tail but most are 40–50 cm long. There is distinct sexual dimorphism in the species, with male platypuses being larger than females. However, there is considerable size variation within a population and among animals from different localities. In general, platypuses from north Queensland are the smallest, with an increase in size going south to Tasmania, where the largest animals are found. A Tasmanian male platypus, weighing in at 3 kg, is three times as heavy as an average male from north Queensland. The species also exhibits size variations that do not appear to be related to its latitudinal distribution.

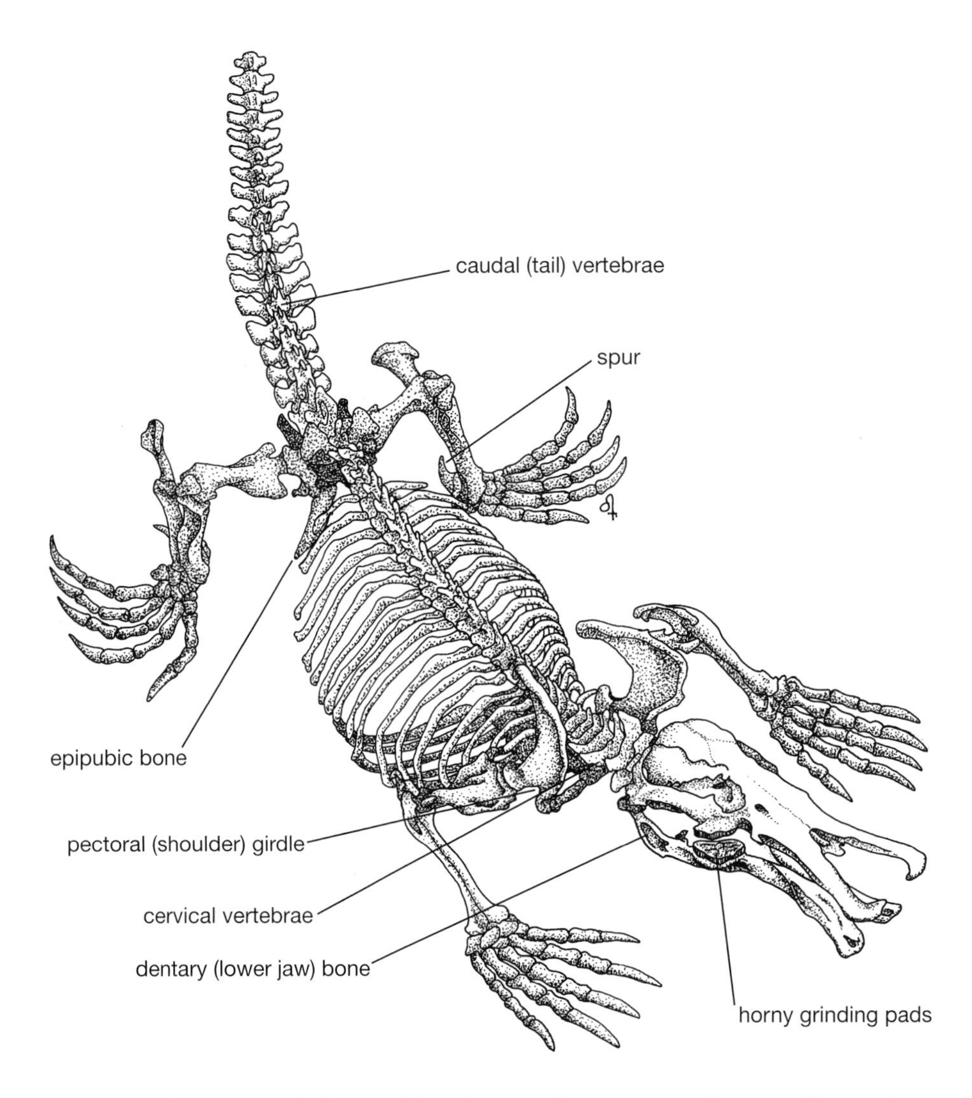

Figure 1.4 The internal skeleton of the platypus shows mammalian, reptilian and uniquely platypus features.

For example, individuals from the west-flowing rivers in New South Wales are larger than those found in the state's east-flowing streams. Table 1.1 shows the weights of samples of platypuses caught at a number of locations by research workers in Queensland, New South Wales, Victoria and Tasmania.

Table 1.1. Weights (in grams) of samples of platypuses from study sites in north Queensland, south-eastern and south-western New South Wales, Victoria and Tasmania (data collected from various research workers).

Location	Male			Female		
	Mean	Range	Individuals	Mean	Range	Individuals
Barron River, north Queensland	1094 ±163	830–1500	n = 6	691±42	610–760	n = 9
Barnard River, northern NSW	1222 ±141	980–1610	n = 21	794±74	660–950	n = 23
Shoalhaven River, south-eastern NSW	1462 ±218	1100–2000	n = 50	908±85	707–1110	n = 50
Murrumbidgee River, south-western NSW	1764 ±243	1260–2460	n = 32	1234 ±92	960–1465	n = 56
Goulburn River, Victoria	1970 ±40	1400–2600	n = 62	1320 ±20	1000–1550	n = 58
Brumbys Creek, Tasmania	2495 ±281	1750–3000	n = 22	1488 ±174	1200–1750	n = 12

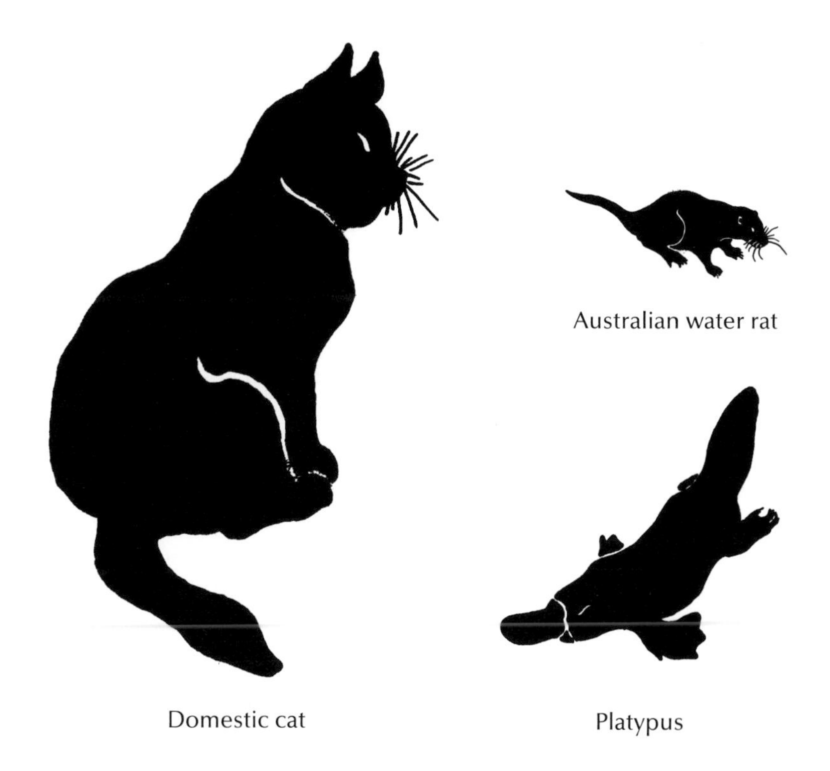

Figure 1.5 The silhouettes of the platypus, the Australian water rat and a household cat show that the platypus is much smaller than many people expect.

The platypus is covered with a dense waterproof fur, except on its feet and bill. It has a streamlined shape, short limbs and propels itself through the water using alternate kicks of its webbed front limbs. Most other aquatic mammals, including the native Australian water rat (*Hydromys chrysogaster*), use their back feet for propulsion. When the animal is walking or burrowing, the webs of the forefeet fold back to expose strong claws. The rear feet are only partially webbed, acting as rudders during swimming. The sharp claws on the back feet are used for grooming and to anchor the body during burrowing activities. Each of the male's rear legs bears a horny spur on the ankle of about 1.5 cm in length. This hollow spur is connected by a duct to a venom gland in the upper leg. A series of vertebrae with flattened transverse processes supports the tail, which consists mainly of fatty tissue and acts as a fat storage area.

The skeleton is streamlined but at the same time heavy enough to support large muscles, particularly in the shoulders and front limbs used both for swimming and digging. In most marsupial and eutherian mammals the pectoral (shoulder) girdle is made up of only two pairs of bones – the scapulae (shoulder blades) and clavicles (collar bones). Interestingly, the pectoral girdles of the monotremes are most like those found in the group of fossil reptiles thought to have given rise to the mammals during evolution. Like the monotremes, these had pectoral girdles formed from five bones, including an interclavicle which, in the platypus, is a distinctive T-shaped bone between the collar bones and the sternum. The interclavicle is also found in modern reptiles. Like the marsupials, the platypus has two epipubic bones attached to its pelvic girdle. The function of these is completely unknown in either group. Suggestions that they support the pouch have been rejected, because the platypus as well as all male marsupials and the females of certain marsupial species do not have pouches, but all have epipubic bones.

The legs of the platypus are splayed, rather like those of reptiles, but they rotate in their sockets as do those of all mammals. The ribs are somewhat reptilian. There are rudimentary ribs on the cervical (neck) vertebrae, as in certain reptiles, but there are seven of these vertebrae, which is a unique mammalian characteristic, like the possession of only a single pair of bones in the lower jaw. In platypuses, the deciduous juvenile teeth are lost a short time after the young leave the nesting burrows, and are replaced by horny pads made of keratin. These 'tooth substitutes' are renewed as the chewing of food and material from the river bed wears them down. Before being lost, the juvenile teeth are hardened with an enamel similar in structure and composition to that found in fossilised adult teeth from one of the extinct platypus species, *Obdurodon insignis* (Chapter 7).

Distribution

Because the platypus is dependent on water in which to feed, its distribution is restricted to parts of Australia where permanent streams occur, although it can persist in occasionally ephemeral streams where permanent pools act as refuges during times when stream flows are reduced or absent due to drought conditions. The overall distribution map shown in Figure 1.6 indicates only the general area of continental mainland Australia, Tasmania and two off-shore islands where the species is known to occupy bodies of water and their riparian margins (stream banks and edges). Even within the catchments of the major rivers, distribution of the platypus may be restricted or fragmented as a result of natural or human influence on the availability of suitable habitat.

Figure 1.6 The current distribution of the platypus is shown in dark shading. There is a small introduced population at the western end of Kangaroo Island. Light shading indicates areas of sparse distribution, absence of records or presence of transient animals. While being found in other coastal catchments in the state, the species does not appear to be resident in the small Portland River system on the far south-western Victorian coast.

The platypus is found in many streams of eastern Queensland and New South Wales and in eastern, central and south-western Victoria. It appears to have always been an uncommon occupant of the rivers flowing across the inland plains in New South Wales and Victoria, although its distribution extends further west along the lowland reaches of the main channels and tributaries of the Murray and Murrumbidgee Rivers than in other west-flowing streams. In Queensland the platypus is found west of the Great Dividing Range in the south but seems to be confined to the eastern side in the central and northern parts of the state. A suggestion that its distribution in north Queensland might be discontinuous east of the dividing range between the Rockhampton and Townsville areas has arisen from recent survey data, but this has not yet been confirmed. The most northerly limit of its distribution appears to be around the Cooktown area and it probably has never included the Cape York Peninsula or the rivers draining into the Gulf of Carpentaria.

Distributed throughout Tasmania, the platypus is also found in a few streams on King Island. In South Australia, individuals have been recorded in the lower Murray River, especially in the Riverland area, but such reports are uncommon and may represent transient non-breeding animals from the New South Wales/Victorian sections of the river. The species once occurred in streams of the Adelaide Hills and Mount Lofty Ranges but now appears to be gone from these areas, where it may never have been common. Translocation of a number of animals from Tasmania and Victoria between the late 1920s and early 1940s resulted in the establishment of populations in several streams at the western end of Kangaroo Island off the coast of South Australia. There is no historical or palaeontological evidence that platypuses ever occurred naturally in Western Australia, although a pair was introduced to an artificial lake south-east of Perth by David Fleay around 1940. There may have been at least two other releases of a few individuals, one possibly in 1951, and there have been reputed sightings of platypuses at several sites from the south of the state to just north of Perth. The latest was in April 2006, but none has been substantiated by a specimen or a photograph.

Platypuses are still recorded from within the greater metropolitan areas of Brisbane, Sydney, Melbourne and Hobart but these distributions are fragmentary and the animals probably occur in lower abundance than they did prior to urban development. Apart from South Australia, where the platypus is considered endangered, an irregular visitor or even extinct, the species is still regarded as common or not given a specific status in the other states in which it occurs. The IUCN (International Union for the Conservation of Nature and Natural Resources) also lists the platypus as being a species of 'least concern'. In spite of this, its dependence on streams and rivers, which are subject to considerable human interference, make the species at least potentially vulnerable to extinction in the long term. There are already documented instances of local extinction and fragmentation of its distribution in certain catchments (Chapter 8).

However, mainly due to its secretive nature, rather than to its rarity, many Australians have not seen a platypus in the wild. *Ornithorhynchus anatinus* has been known to Europeans and to science for more than 200 years and to indigenous Australians for much, much longer. Despite this, the platypus story is far from complete.

The following chapters present what is known about the biology of the platypus. The most unique features of the animal are, of course, its oviparity (egg-laying), venomous spurs and its major sensory organ being its bill, rather than its eyes, ears or nose. These aspects of its biology are discussed in the following three chapters (Chapters 2–4). Its foraging, diving and energetics are discussed in Chapter 5 and other aspects of its ecology in Chapter 6. Its evolution and ancestry are dealt with in Chapter 7, before Chapter 8 discusses its conservation, including the interactions between the platypus and humans. In most of these chapters some thoughts and hypotheses are also offered regarding gaps in knowledge of its biology and in Chapter 9 a series of questions (FAQ) arising from the general subject areas in the other chapters is presented.

2
BREEDING BIOLOGY

'Monotremes oviparous, ovum meroblastic'

WILLIAM CALDWELL (1884)
Message to the British Association for the Advancement of Science meeting in Montreal, announcing that the platypus and echidna did in fact lay eggs.

Following the first report of the platypus in 1798 by Governor John Hunter, the answer to the question 'Do platypuses and echidnas give birth to live young or lay eggs?' soon became the 'holy grail' of a number of biologists in Britain and Europe and especially many of those who visited the Australian colonies. To resolve this question – the answer to which had almost certainly been known by Aboriginal people for millennia and already tentatively reported by other Europeans – the Scottish biologist William Caldwell slaughtered many hundreds of platypuses and echidnas during his work in Australia. In 1884, with the evidence in hand, he sent a succinct message – *'Monotremes oviparous, ovum meroblastic'* – to the British Association for the Advancement of Science. He had at last acquired convincing evidence that not only did platypuses and echidnas lay eggs but that the embryo develops by partial cleavage of the egg, as occurs in birds and reptiles, rather than by complete cleavage as in marsupial and eutherian mammals. Interestingly, a representative of TW Crawley and Company of Hunter Street in Sydney took exception to Caldwell's report at the time, claiming in a letter to the *Sydney Morning Herald* on 20 September, 1884, that their hunters had 'dissected a great number' and taken platypuses from burrows but had found no eggs either inside the animals or in their burrows. They proposed instead that platypuses spawned eggs like frogs but in smaller numbers!

Reproductive system

All species of monotremes possess the mammalian features of hair and mammary glands, but have characteristics that distinguish them from other mammals. Females lay eggs but have no teats. Males have testes housed inside the abdomen (testicond), rather than in a scrotum, and produce filiform (thread-shaped) reptile-like sperm cells. Both sexes also have a structure called a cloaca – a single external opening into which the reproductive, excretory and digestive systems open.

The female reproductive system in the platypus opens into the cloaca via an ante-chamber called the urogenital sinus. This sinus is connected to separate left and right reproductive tracts, each comprising an ovary, oviduct and uterus (Figure 2.1a). As in many bird and some reptile species, only one side of the female reproductive system is functional (left side), but despite this functional loss of one ovary, the platypus usually ovulates two eggs, and sometimes three, during the breeding season. By contrast, in female echidnas both ovaries are functional but only one egg is normally laid each breeding season.

In the male platypus, each abdominal testis is connected by a duct to a large secretory epididymis (structure where the spermatozoa mature). This duct, the vas deferens, conveys semen through the urethra of the penis during copulation. Urine is voided from the bladder into the urogenital sinus from where it passes to the outside of the body through the cloaca (Figure 2.1b).

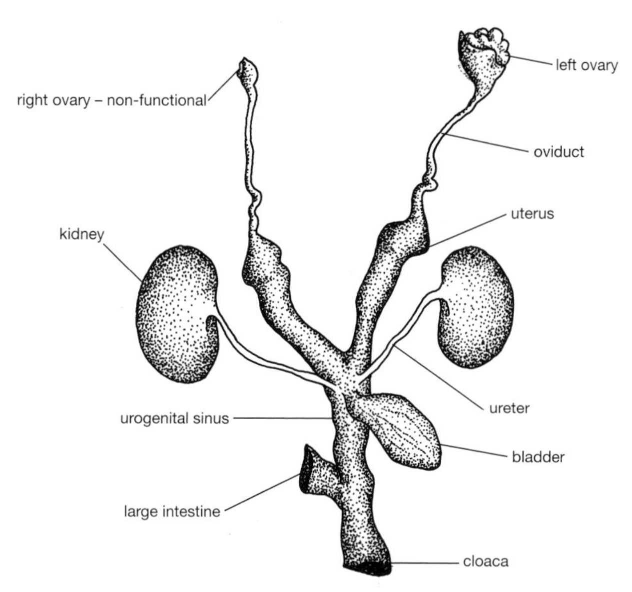

Figure 2.1a In all monotremes the reproductive, excretory and digestive systems open into a single chamber, the cloaca, which has one external opening. (The word 'monotreme' means 'one hole'.) In the female platypus only the left ovary is functional.

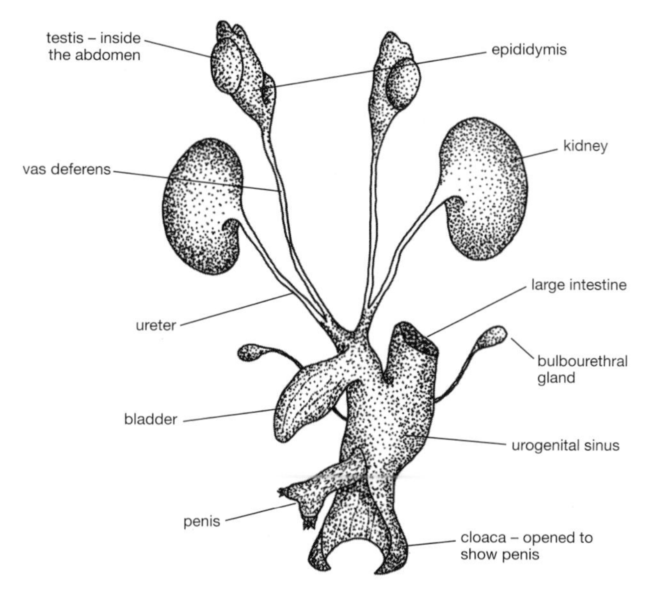

Figure 2.1b In the male platypus the testes are inside the abdomen, rather than in an external scrotum. The ureters from the kidneys and the ducts from the testes open into the urogenital sinus but urine is voided from the cloacal opening and only semen is passed through the urethra of the penis.

Breeding season

The reproductive organs of male and female platypuses in the tableland areas of New South Wales and Victoria begin to increase in size around late June, and reach their maximum in August to October when mating normally occurs. After this time a regression of these organs to non-breeding condition occurs. As with other mammals, these changes are brought about by hormonal variations in the platypus.

Sperm production in the male is controlled by gonadotrophic hormones (follicle stimulating hormone [FSH] and luteinising hormone [LH]) from the pituitary gland and by androgens (including testosterone) secreted from the testis. Levels of androgens in the blood rise considerably during the breeding season, causing the testes, epididymides and the venom glands to increase significantly in size (Figure 2.2). The testes of a non-breeding male platypus together weigh a little over one gram, while those of a breeding animal can be as heavy as 22 grams; an increase of around 20 times in size. At this time males also become more aggressive. In the female the hormone progesterone, which is involved with maintaining the lining of the uterus during the short pregnancy, also increases

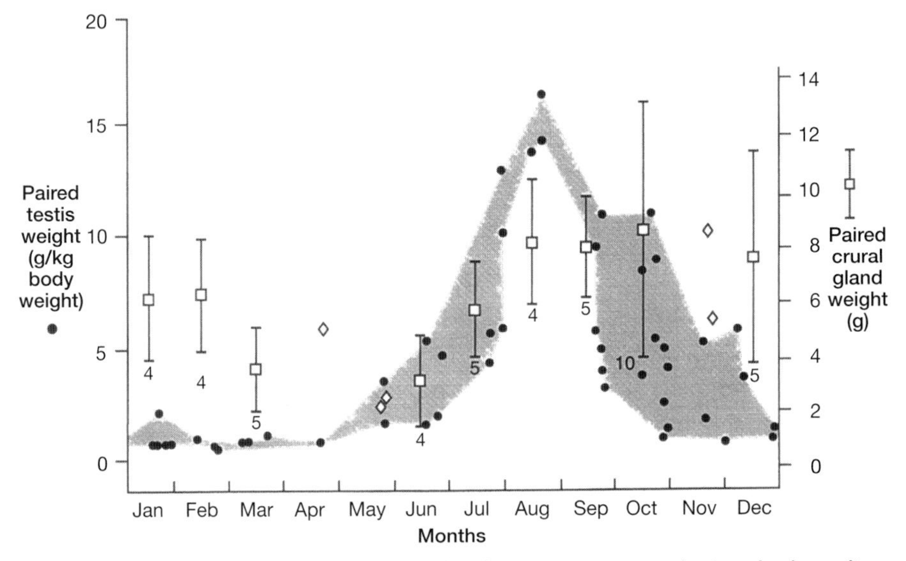

Figure 2.2 The size of the testes of the male platypus increases during the breeding season. Black circles show mean (average) values of testis weight per kilogram of body weight for each month, and the shading indicates variation (standard deviation). The venom glands also increase in size. White squares show mean gland weights and bars standard deviations. Figures for each month represent the sample size. Diagram: Peter Temple-Smith

to reach peak levels during August and September, before declining during October. This ensures that nutrients are passed from secretory glands in the uterine walls of the mother to the embryo developing inside the unlaid egg. It is not known what initiates these hormonal changes, but in other animals seasonal environmental changes are involved in the timing of breeding.

Fragmentary observations, inferences from female lactation and times when juveniles appear in the streams all suggest that breeding occurs earliest in Queensland, later in New South Wales and Victoria and later again in Tasmania. In New South Wales, platypuses normally mate around August to October and lactating females are found from late September to early March. In Tasmania on the other hand, a lactating female was found in mid-May at Sisters Creek in northern Tasmania, juveniles were found entering the population in Lake Lea from April to June, and in a small sample of male platypuses collected from a number of Tasmanian streams, the maximum testis size was recorded in November, compared to August–September in New South Wales (Figure 2.2). Even within individual platypus populations at different geographical locations there is a variation in breeding times: some females mate several weeks earlier than others in the same population. As a result, juveniles can enter the population over a period of some months in each location, beginning in January in the north and as late as June in Tasmania.

As an example of the timing of breeding events in the platypus, a mating and fertilisation in a stream on the central tablelands of New South Wales on 11 September would result in eggs being laid around 1 October (21 days gestation), hatching on the 12th of that month (10 days incubation) and the young emerging from the nesting burrow in early February (after four months of suckling). These events are outlined in Figure 2.3.

Reproductive behaviour

Only fragmentary and occasional observations of platypus breeding behaviour have been made in the wild due to the secretive, mostly nocturnal, habits of the species. Most observations of platypus behaviour have been made of captive animals which show a considerable repertoire of interactions during the mating season. Copulation has been observed a number of times in captive animals held in various Australian zoos but

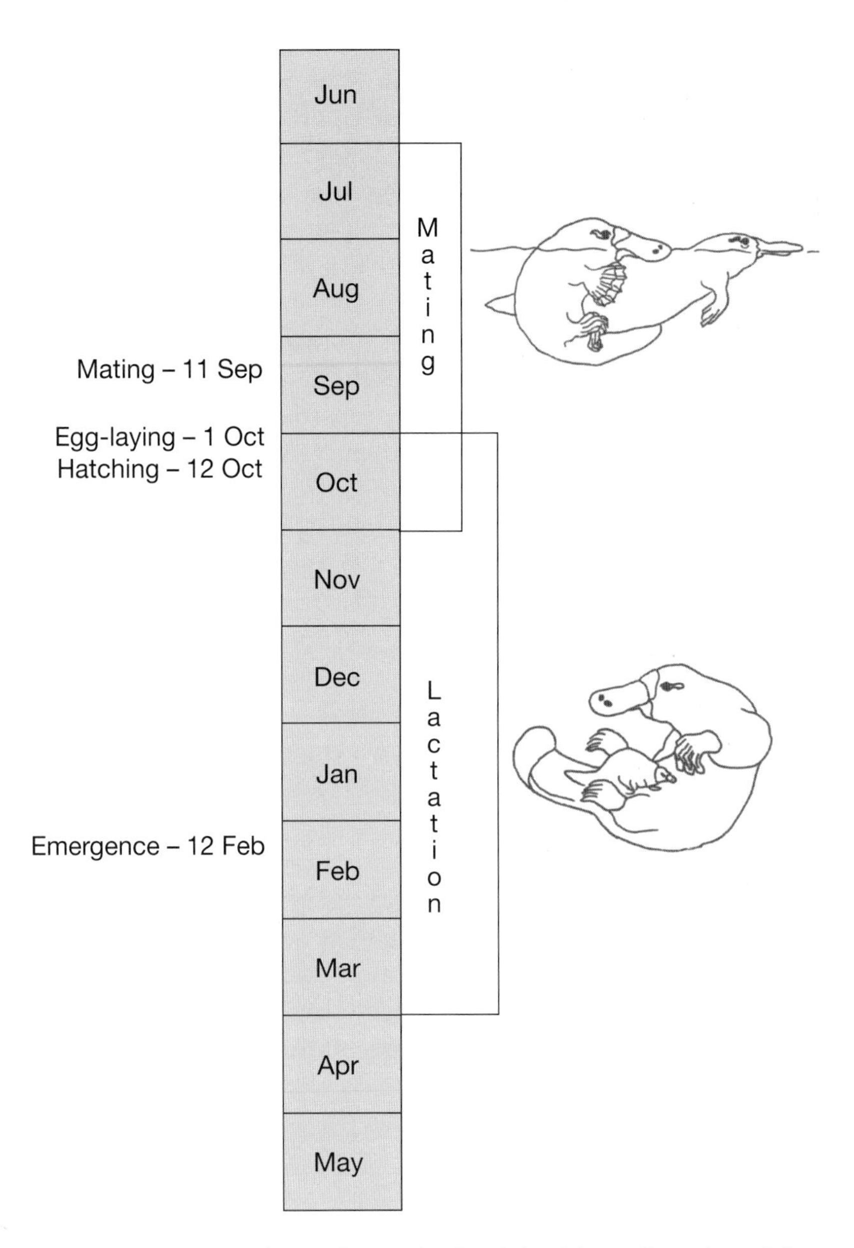

Figure 2.3 Mating occurs earlier in the north of mainland Australia and much later in Tasmania, so that juveniles enter the populations at different times depending on geographical location. There is also variation in the timing of reproduction within individual populations. The figure shows the timing of events during the breeding season in the upper Shoalhaven River in New South Wales. A successful mating in the middle of September would result in the young emerging in mid-February.

very seldom in wild animals. Males initiate most mating interactions, but the compliance of the female appears to determine if pre-mating behaviour culminates in copulation. Pre-copulatory behaviour includes a range of courtship activities that last from less than a minute to half an hour or more, usually occurring over several days. These include animals grasping each other and rolling sideways several times, diving, touching and passing. Often such interactions are broken off and recommenced. The male, usually recognised by its larger size, is frequently seen grasping the tip of the female's tail, after which both animals swim in tight circles. During a typical copulation, the male grasps the female by the tail from behind with its bill. He then wraps his tail under the female's body to one side of her tail, and as he moves forward he nuzzles and grips the fur on her shoulder or neck with his bill. With their bodies in this position the male everts his penis through the cloacal opening and inserts it, via the female's cloaca, into her urogenital sinus where the sperm begin their journey to fertilise the egg in the oviduct. In all instances observed in captivity, a number of copulations occurred over several days. The nature of certain interactions and copulation are illustrated in Figure 2.4.

Only the female builds the nest. In captivity she spends four to five days renovating the burrow and for a similar period of time she collects wet nesting material, which is held between the rear feet and the curled tail while being dragged to the burrow. This wet vegetation is thought to provide a level of humidity required to prevent desiccation of the eggs or nestlings during incubation and after hatching. Several observations of female platypuses carrying nest material have also been made in the wild and excavated nests have contained mixtures of plant parts, which usually reflect the surrounding riparian vegetation.

Over much of their distribution it is in winter, or at least when water temperatures are still quite low, that platypuses enter the water to engage in reproductive behaviour. This is also the time when energetic demands are highest (Chapter 5). In Victoria, a study of the blood levels of adrenal gland hormones involved in making stored energy available for activity (glucocorticoids) found these hormone levels much higher in both males and females during the period when mating behaviour would have been occurring in that region. Figure 2.5 shows blood being collected from a platypus for hormone analysis. Although it was not known whether the individuals captured were involved in these breeding activities, this observation indicates there is considerable energy demand at this time of the year that may be associated with reproductive behaviours, possibly even involving burrowing and nest-building in females.

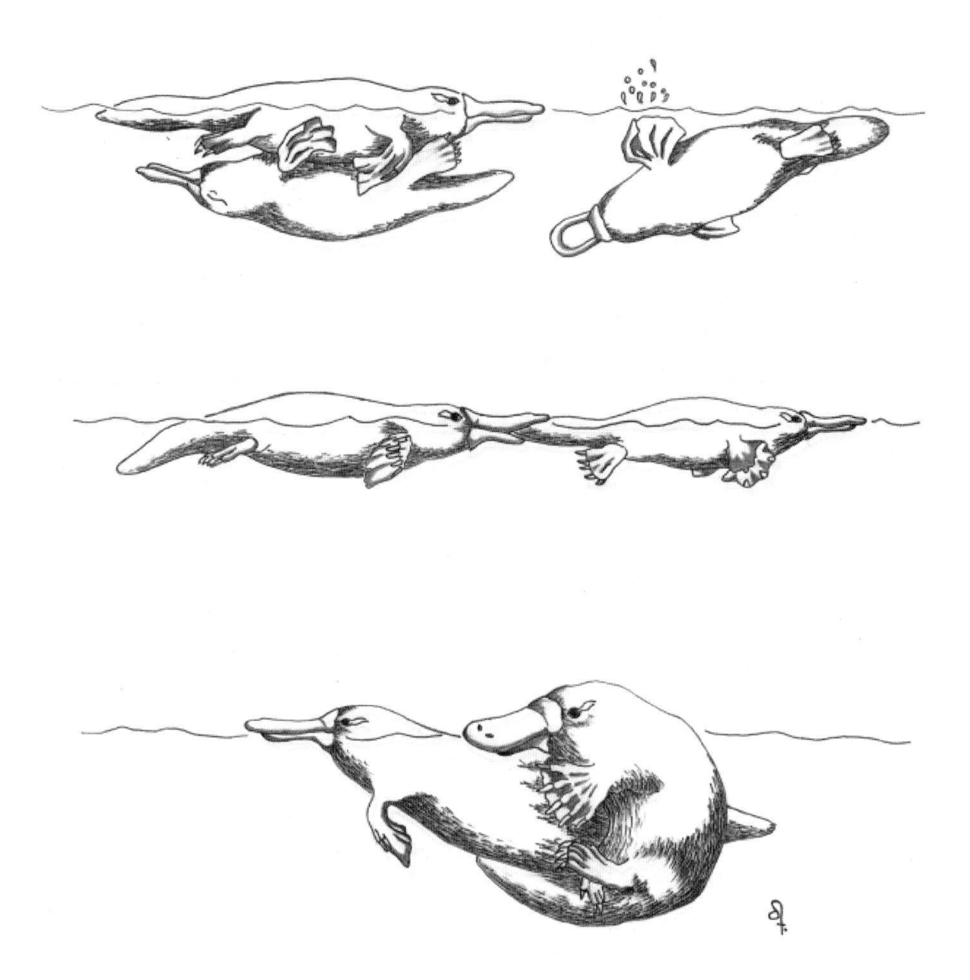

Figure 2.4 Mating behaviour in the platypus involves touching, rolling, diving and grasping by the tail, sometimes followed by copulation.

From fertilisation to hatching

Platypus eggs develop from oocytes inside ovarian follicles only in the left ovary, the right being non-functional. Some of these follicles increase in size in spring and produce a hormone (oestrogen) that stimulates the enlargement of the two uteri. When it is ovulated from the follicle, the egg is about 4 mm in diameter and is fertilised by the sperm in the oviduct. The sex of the embryo is determined at this time.

Sex determination in the platypus is very unusual indeed. As in other mammals, it is determined by the inheritance of X and Y chromosomes,

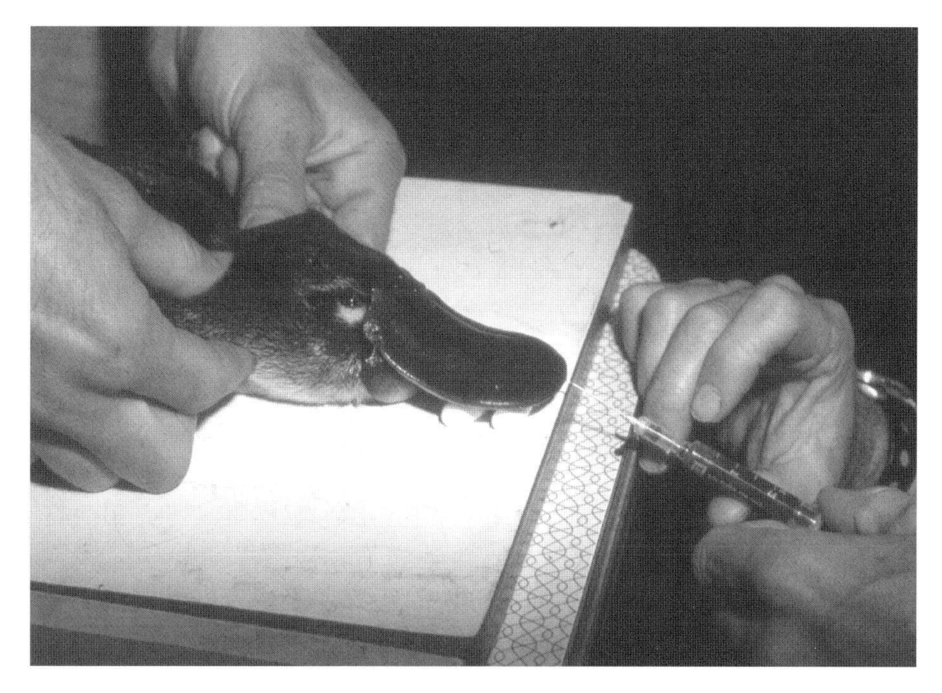

Figure 2.5 A small blood sample (up to 2 mL) can be taken from a large blood vessel in the front of the bill. The procedure is actually less painful than it looks and has animal ethics approval. Photo: Richard Whittington

but rather than having one X and one Y chromosome, the male platypus has five X chromosomes and five Y chromosomes. The female has five pairs of X chromosomes. During meiosis, leading to the production of sperm, the five X and five Y chromosomes duplicate then segregate as separate chains, as do the two sets of five X chromosomes during meiosis in the female. There are geneticists who suggest that the platypus sex chromosomes may be derived from a slightly different sex determination system found in birds and reptiles, rather than the usual XY system found in other mammals.

After fertilisation the first shell layer is laid down and the egg (or eggs) passes down the oviduct into the uterus where the second and third layers of the shell, secreted by glands in the walls of the uterus, are added. The uterus also supplies nutrients to the egg, facilitating its increase in size to about 14 mm in diameter by 17 mm in length by the time it is laid (see Colour Plate (top left), page 34). The gestation period from fertilisation to egg-laying is about three weeks, which is similar to the 23-day gestation period reported for the short-beaked echidna.

It is thought that the egg is laid directly onto the abdomen of the mother, where she incubates it by adopting a curled posture, with her tail laid over the underside of her body and the tip covering her bill (Figure 2.6). Platypuses also adopt this posture while sleeping, as it conserves body heat by covering most unfurred surfaces, including the front feet and most of the bill, with furred parts of the body. Development of the embryo occurs during incubation, again of unknown length, but thought to be about 10 days. At the end of this period the young breaks through the parchment-like shell of the egg and begins to take milk from its mother. The dimensions of the short-beaked echidna's egg are almost identical to that of the platypus and the embryo at laying has developed to a similar stage. A newly hatched platypus has not yet been measured but is probably similar in size to that of the echidna, which is only 15 cm long (see Colour Plate (top right), page 34). Up to three eggs can be laid by the platypus, but it is not known how many of these hatch or the number of hatchlings that are successfully reared. Breeding in captivity has resulted in only one or two nestlings emerging from the nesting burrow as juveniles.

Egg-laying, incubation and hatching often occur in very long complex burrows, with a number of passages and chambers constructed and possibly re-worked from season to season by one or more females (Figure 2.7). Such burrows can measure up to 30 metres in length but nesting may occur in much simpler burrows of only a few metres. Nesting burrows have almost exclusively been found in earth banks. However, at Lake Lea

Figure 2.6 Platypuses assume this curled posture when sleeping, which reduces heat loss from unfurred parts of the body by tucking them into furred areas. Certain observations suggest that the female incubates the eggs by curling the body around them in this way.

Figure 2.7 This excavated nesting burrow shows a long tunnel (which can be up to 30 metres long) and several chambers, one of which is lined with nesting material.

in the Cradle Mountain area of Tasmania, a lactating female was radio-tracked to a burrow located at the base of shrubs (*Richea scoparium*) growing along one of the creeks draining into the lake. Platypuses are occasionally reported foraging and sheltering in caves, both on the mainland and in Tasmania. Recently, two accumulations of nesting material and platypus hair were found 160–200 metres inside a Tasmanian cave, which also suggests that at least some individuals are considerably flexible when it comes to their selection of nesting sites.

Observations of captive breeding have challenged long-held assumptions about platypus breeding biology and behaviour. It seemed inconceivable, once incubation had commenced and prior to hatching, that the female would be able to leave the eggs. Observations from the 1943/44 breeding at the Sir Colin Mackenzie Sanctuary at Healesville in Victoria, however, suggested that about six days into the incubation period the female exited the nesting burrow and returned to the water for a brief time before continuing the incubation. Data from more recent captive breeding, at both the Healesville Sanctuary in Victoria and Taronga Zoo in Sydney, showed that females left the nest for less than an hour after as little as 54 hours (2.25 days) from first retiring to the nesting burrow. Although extended periods were spent in the burrows, females emerged either on consecutive days or intermittently during the presumed 10 days of incubation.

It is completely unknown where the female deposits the eggs or early hatchlings during these early excursions from the nesting burrow and how, on her return, she manages to reposition the eggs or young to continue brooding them. In his description of finding monotreme eggs, which contributed to his announcement on oviparity, William Caldwell reported obtaining the platypus egg from a female he had shot on 24 August 1884. He observed that it had already laid one of its eggs but still had the other in its uterus. Assuming that he dispatched the unfortunate animal while it was foraging, rather than shooting it after it had been unearthed from a burrow, the female must have left its nesting burrow between laying one egg and the impending deposition of the other. It is unclear how Caldwell determined that another egg had been previously laid, but clearly, more than a century after his historic announcement to the scientific world, a full understanding of platypus reproduction has not been achieved. With the recently improved breeding success of the species in captivity and the availability of modern optical surveillance and recording techniques, detailed observations of events inside the nesting burrow are eagerly anticipated.

Male and female platypuses do not appear to reproduce during the breeding season immediately following their emergence from the nesting burrow. Both sexes are capable of breeding in their second year, although many females in the upper Shoalhaven River in New South Wales were found not to do so until they were over four years old. Mechanisms controlling when females breed and which individuals in a population do so are not known. However, at least one juvenile female released from Kangaroo Island into Warrawong Sanctuary in the Adelaide Hills produced young in her second breeding season. It is also reported that three young were produced in the second breeding season after several juveniles from a nearby stream had been released into Cardinia Creek in Victoria, where the resident population was reported to have died out after the 'Ash Wednesday' fires of 1983. These examples suggest that more females may successfully reproduce at a younger age in areas where population pressures are lower and where there may be less competition from established breeding females.

Mammary glands and milk

In the mammary glands of the platypus, special cells produce milk components. The milk is like that of other mammals, containing casein molecules (the main milk proteins), whey proteins, carbohydrates, fats and various minerals. Most mammals have at least two types of casein molecules in their milk. Echidna milk contains two caseins, but platypus milk has only one. Biochemical analysis of the milk has shown it to be very rich, containing more total solids than that of many other mammals. Platypus, echidna and marsupial milk all have high concentrations of iron, necessary for the formation of haemoglobin in red blood cells. The young of eutherian mammals have iron stored in their livers when they are born, but because the young of the monotremes and marsupials are so small, their livers cannot store enough iron and it must be supplied in the milk during the suckling period. The fat composition of the milk is largely determined by diet, and the fatty acids found in platypus milk are those found in the macroinvertebrate food species on which it feeds (Chapter 5). As in marsupials, there are large changes in the volume and composition of platypus milk during lactation.

Milk is secreted into a network of ducts in the mammary glands that collect together at the surface of the skin in two milk patches, or areolae (singular: areola). These patches are very similar in structure to the nipple area in humans but do not protrude and are covered with fur. Prodding by the young platypuses at these areas is thought to cause milk to be ejected onto the fur, just as the sucking of a human baby produces milk flow from the mother's nipple. This response is mediated through the pituitary gland at the base of the brain, which produces the hormone oxytocin in response to the sucking (or prodding in the case of the platypus) by the young. Oxytocin is carried in the bloodstream and stimulates the contraction of a network of special muscle cells in the mammary glands which forces the milk out through the areolae.

From May until July, the mammary glands of the platypus each consist of a small structure, only about 1 cm in length, under the skin on the female's abdomen. Towards the end of this period, these glands begin to enlarge and develop into large fan-shaped structures, occupying much of the ventral abdominal surface of the body and may even extend up towards the back (dorsal surface). The glands probably enlarge in most females at this time but do not begin to produce milk until after hatching of the young has taken place. In field studies, by injecting a small dose of the hormone oxytocin, milk can be expressed from the glands if they are gently squeezed. This is done to identify females with young during the breeding season.

Inside the nesting burrow

The young platypus spends almost its entire first summer being suckled in the nesting burrow. It does not see the light of day, or the river into which it will become independent, until three to four months after it hatches from the egg. In captivity, lactating females initially spend most of their time in the nesting burrow but as the period of lactation progresses, they spend more time in the water feeding. Data from the successful captive breeding at Healesville Sanctuary in the 1998/99 breeding season showed that the lactating female consumed 900 to 1000 grams of food each day during the final stages of suckling her two nestlings (90–100% of her own body weight), indicating the high metabolic demands of lactation. These extended periods of feeding in captivity where food is readily available are intriguing, as wild platypuses must spend time foraging for prey organisms to meet the same demands of lactation. In spite of such an apparently extreme metabolic burden during lactation, one female captured many times in the upper Shoalhaven River in New South Wales was still breeding at 21 years of age.

Prior to lactation, the captive female at Healesville Sanctuary consumed 20–30% of her body weight in food per day and other captive and wild platypuses are known or estimated to consume only 13–28% of their body weight (Chapter 5). This dramatic increase in intake during the latter part of lactation is surprising and requires further investigation, especially in wild animals. The notion that for a few weeks during late lactation, each female platypus requires a daily intake of about 1 kg of small macroinvertebrate prey items seems difficult to accept, especially in some streams, where a number of lactating females may occur in a particular area and other species, including water rats, water birds and native and introduced fish species, also compete for this food. If, for example, three lactating females of around 1 kg in weight, three non-lactating females of about the same weight and two males of 1.5 kg use a section of a stream, more than 4 kg of macroinvertebrate organisms would have to be harvested each day to satisfy the dietary requirements of the platypuses alone. Perhaps under field conditions, both lactating and non-lactating animals may be able to survive on less food than was consumed by captive animals in Healesville Sanctuary (Chapter 5). A detailed study of the field condition and dietary intake requirements of lactating and non-lactating platypuses in relation to the distribution, density and availability of macroinvertebrate organisms is necessary. It may be that competition for food by wild platypuses, and the inability of some breeding females to consistently obtain the large quantities of invertebrate organisms needed to maintain lactation, determine the successful reproduction of individual female platypuses in wild populations.

In the upper Shoalhaven River in New South Wales, 80% of adult females captured were found to be lactating in certain years, but in other years fewer than 30% were producing milk. A number of breeding females in the area did not breed in consecutive years. In particular instances apparently low reproductive success could be attributed to obvious environmental constraints, such as drought, but others were unexplained. Perhaps the intense metabolic demands of three to four months of lactation cannot be maintained from year to year if foraging conditions are even slightly less than optimal.

During its time in the nesting burrow, the platypus nestling grows very quickly (see Figure 2.8 and Colour Plate (bottom), page 34). Measurements of preserved museum specimens showed that during the first two and a half weeks after hatching, nestlings increased in size by between three and seven times and were 20 times the hatching size (around 1.5 cm), by 14 weeks of age. The recorded variation in size between individual nestlings is probably attributable to the availability of milk, which would be affected by a number of factors including litter size (one to three young) and food available to lactating females. As well as increasing in length and weight, the dimensions of various parts of the body change as the nestling develops into the young juvenile that will leave the nesting burrow. For example, the nestling's bill is almost as broad as it is wide, while that of the emerging juvenile platypus is 'stubbier' than that of an adult, but longer than it is wide (Figure 2.9).

Life as a juvenile

When the young finally emerge from the nesting burrows they are smaller than adults of the same sex, but males are already significantly larger than females (sexual dimorphism). In the upper Shoalhaven River in New South

Figure 2.8 These two nestlings, inadvertently unearthed from their nesting burrow by workmen, are probably less than two months old. In an area where the platypus occurs, it is advisable that any excavation work close to stream banks should not be done during the breeding season when there might be dependent young in nesting burrows. Photo: Faye Bedford/Victorian Department of Sustainability and Environment

Wales juvenile males average 779 grams in weight and 41.2 cm in length compared to females, which are 588 grams and 37.5 cm in weight and length respectively. In several streams in eastern New South Wales, including the upper Shoalhaven River, newly emerged juveniles were found to be 56–75% of adult weight and 80–88% of adult length. Compared to that of the nestlings, the growth rate in juveniles is less, but by the end of 12 to 18 months from the time they leave the nesting burrows, the young are nevertheless at, or very close to, adult size. By 24 months growth appears to have ceased, although there can be significant weight changes related to season and/or food availability (Chapter 5). By this time, males are not only of greater length than females but also appear more heavily built.

Once they leave the nesting burrows, it is not known if juveniles continue to be fed by their mothers or, if suckling does continue, how long it lasts. During a study in the upper Shoalhaven River in New South Wales, one juvenile captured in late January coughed up a little milk when its

Figure 2.9 The bill of this newly emerged juvenile male, bred in a semi-natural environment at Warrawong Sanctuary in the Adelaide Hills, is longer than it is wide, but it is still stubbier than in adult animals. Photo: Russell Millard

cheek pouches were being sampled for diet analysis (Chapter 5). These cheek pouches contained macroinvertebrate material, showing that the animal was taking solid food and milk at the time it was captured. Data from observations in captivity relating to suckling after the young first emerge from nesting burrows are equivocal but suggest continuation of suckling by emerged juvenile animals that are also foraging for solid food. A number of young were observed going back to the nesting burrow after initially emerging, while others did not return. Sometimes the mother visited the nesting burrow after the young had emerged but it is not known if suckling of the young occurred in these instances. Staff at Healesville Sanctuary reported that mammary gland tissue was physically noticeable on the female involved in the 1998/99 breeding when she was first handled 138 days after the presumed laying date. However, when caught again on day 158, the mammary tissue was no longer evident, suggesting that suckling still occurred for several days at least after the second nestling emerged on day 136. On neither of these occasions, however, was the injection of oxytocin carried out to confirm lactation.

A most intriguing observation in relation to lactation and juvenile behaviour was reported by Jules Verreaux, a trained naturalist from France, who visited the colony of Tasmania from December 1842 to April 1844. His translated account reads as follows:

> *Having near at hand (still on the banks at New Norfolk) a considerable number of adults and juveniles; I saw the young accompanying their mothers with whom they played, especially when they were too far from the bank to feed. I clearly saw that when they wanted to get some food, they took advantage of the moment when the mother was among the water plants, some distance from the land, where there was no current. As the female had her whole back exposed, you could easily work out how once considerable pressure had been applied, the milk floated up a short distance, and the young one could easily absorb it. This manoeuvre is easy to notice, as one can see the bill moving rapidly. The best way I can describe the female's oily liquid is to liken it to the iridescent colours produced by sunrays on stagnant water. I saw the same process repeated every day and night and I also observed that when the young one was tired, it climbed on the back of the mother, which headed for land where the young one stroked her.*

Neither the observations of the young taking milk from the surface of the water nor their travelling on the back of the mother have been reported by other observers in either wild or captive situations since that time. Platypus milk is a pink-white liquid and an experiment showed that the solid fraction in fact sank as a white precipitate in water. A very thin film of fatty material did float on the surface but was quickly dispersed by water movement, such as might be expected to happen around a swimming platypus and her young. No iridescence was observed in dim light or full sunlight. It is not known whether Verreaux's description represents observations made after spending time in one of the many inns that apparently flourished in New Norfolk at the time, local folklore or a serendipitous observation of unique behaviour in this cryptic species not reported before or since.

No comprehensive study has been carried out into the fate of juvenile platypuses once they become independent. In 'good' years, young will probably disperse and take up residence in poorer habitat along tributary streams but in 'bad' years juvenile mortality will probably be high. A study in the upper Shoalhaven River in New South Wales indicated that most juvenile platypuses almost certainly left the local area by midway through their first year of life. Certain juvenile females did remain in their natal area and became part of the breeding population but, of all the juveniles marked and released during the study, only 32% of females and 14% of males were ever captured again in the area. The few juvenile males that were recaptured were predominantly caught again within the first few months after initially emerging from the nesting burrows and then were not captured again.

The social system in platypuses is completely unknown. After emerging from their respective nesting burrows, individuals appear to live solitary lives. There is no evidence of their occurring as 'families' or 'pairs', as is often perceived when several individuals are observed in a single reach of a stream. The retention of the crural system (spurs plus their associated venom glands) only in male platypuses would seem to indicate that the crural system is associated with territorial and breeding activities. This notion and the nature and function of the spurs and associated venom apparatus are considered in detail in the following chapter.

Swimming on the surface, the platypus creates a characteristic bow wave. This animal is alert with both its ears and eyes open. Photo: © Ian Montgomery, birdway.com.au

As the platypus dives, water pressure squeezes air from its fur. The eyes and ears are tightly closed during diving. Photo: © Reg Morrison/AUSCAPE

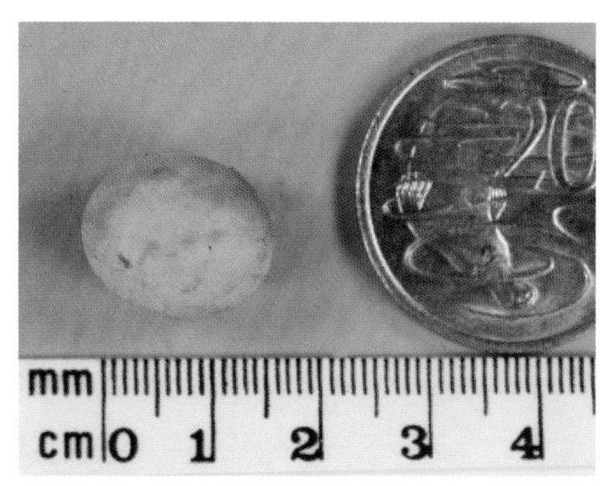

It was nearly 90 years after the first platypus was reported by Governor Hunter that the first platypus egg was seen by a European.

A platypus hatching from an egg has not been seen but it would be in a very similar state of development and size to this hatchling short-beaked echidna. Photo: Ederic Slater

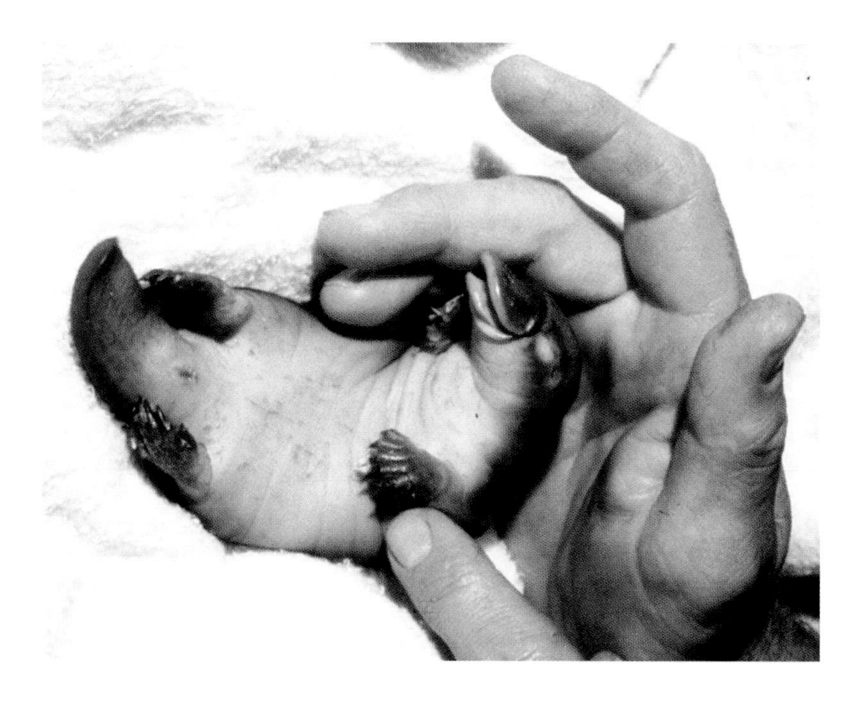

Its eyes still closed and incompletely furred, this nestling platypus is probably only a few weeks old. It had apparently left the nesting burrow near a dry creek during drought conditions on the central tablelands of New South Wales and was sent to Taronga Zoo in Sydney.

The venomous spur on each hind leg of the male platypus is about the size of a large rose thorn. In this picture small juvenile platypus ticks (*Ixodes ornithorhnchi*) can be seen around the rear ankle of the platypus.

A few large adult female platypus ticks can be found in the body fur and attached to the front ankle of platypuses. Photo: © David Parer/AUSCAPE

The Thredbo River – winter 1981. Frost and ice are not unusual in areas occupied by the platypus in the highlands of Tasmania, Victoria and New South Wales.

Because of its excellent insulating ability, platypus fur was sought after for blankets, gloves and hats, leading to the slaughter of thousands of animals until the species' protection around the turn of the 20th century. Photo: Peter Aitchison/Australian Geographic

This unfortunate platypus was found hooked and drowned on an illegally set fishing line in the headwaters of the Ben Chifley Dam in New South Wales. Photo: Matt Ryan

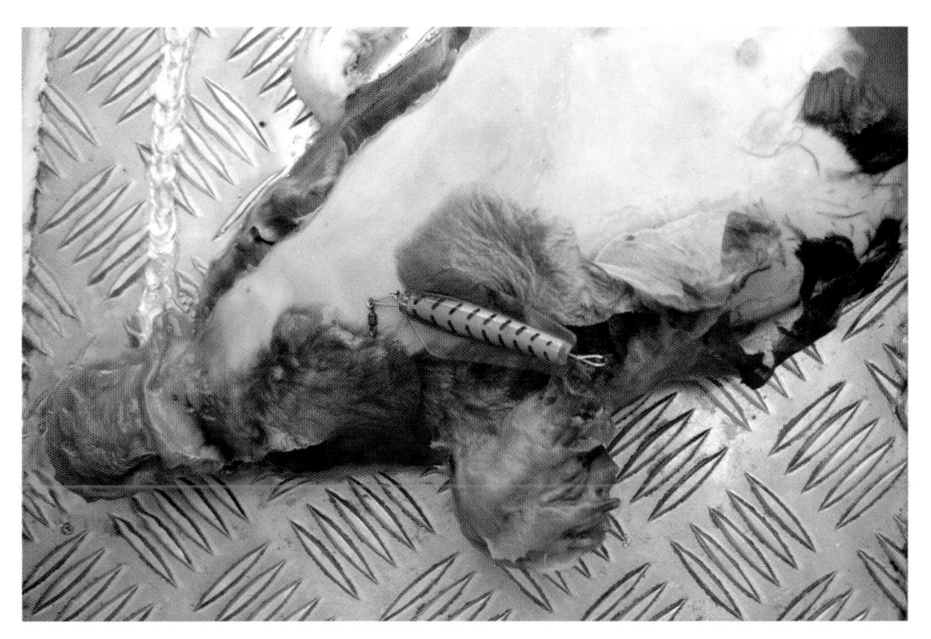

Caught by a 'Tassie Devil' fishing lure, this partly decomposed platypus was found floating in Lake Jindabyne in New South Wales. Photo: Matt Ryan

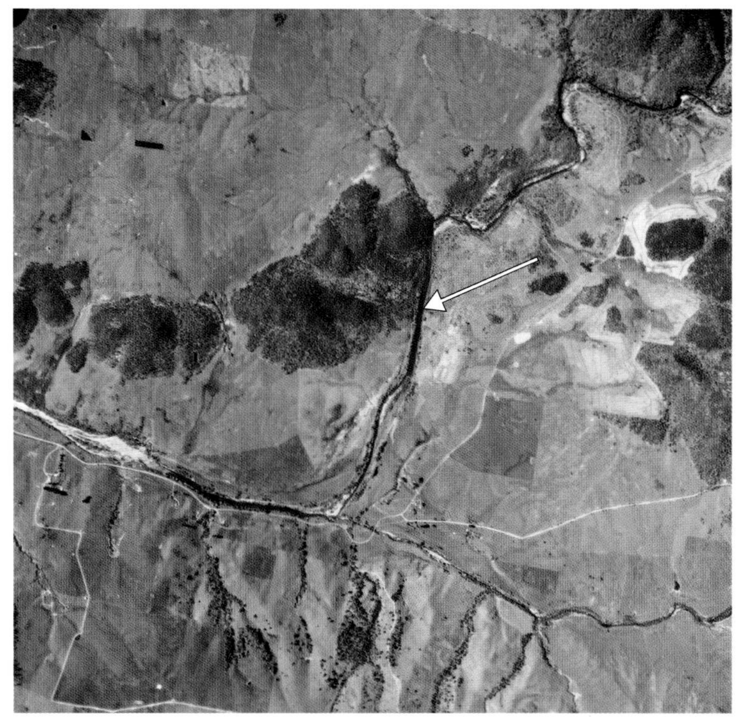

In 1977 this large pool on the upper Shoalhaven River had an average depth of 2 metres and was the preferred foraging and breeding area for many platypuses.

Aerial photograph: © NSW Department of Lands, www.lands.nsw.gov.au

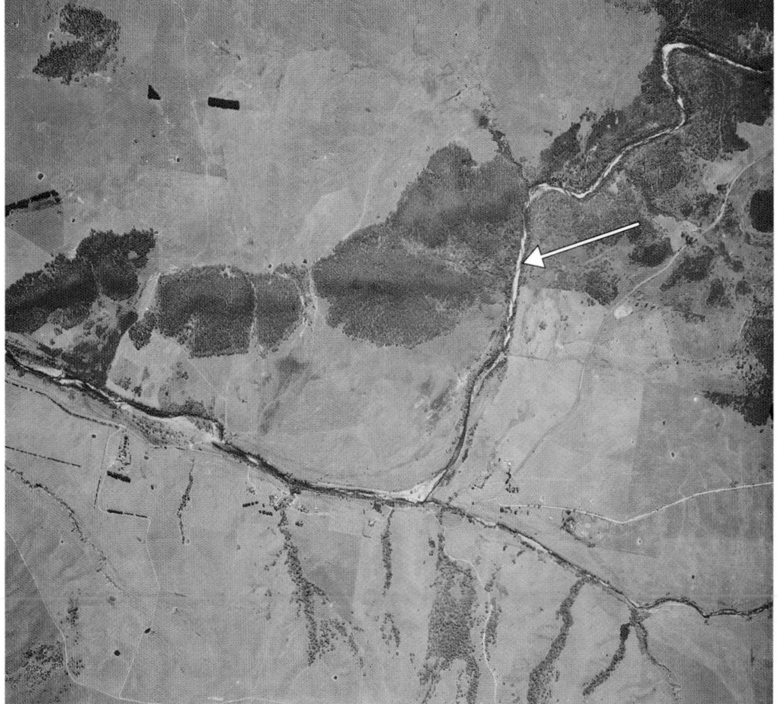

By 1997, due to erosion brought about by clearing of riparian vegetation, cattle access, wombat activity and several floods, the pool had filled up with sand and was no more than knee-deep. Platypuses now seldom use it. The lack of plant cover seen in the aerial photograph resulted from the effects of drought conditions in the area. Aerial photograph: © NSW Department of Lands, www.lands.nsw.gov.au

Platypuses normally live in burrows with entrances close to the water level but some have been found sheltering in accumulated river debris (top). The disturbance of small sticks at the mouth of the burrow (bottom) indicated that it was in use.

Photo: © Ian Montgomery, birdway.com.au

3

SPURS AND VENOM GLANDS

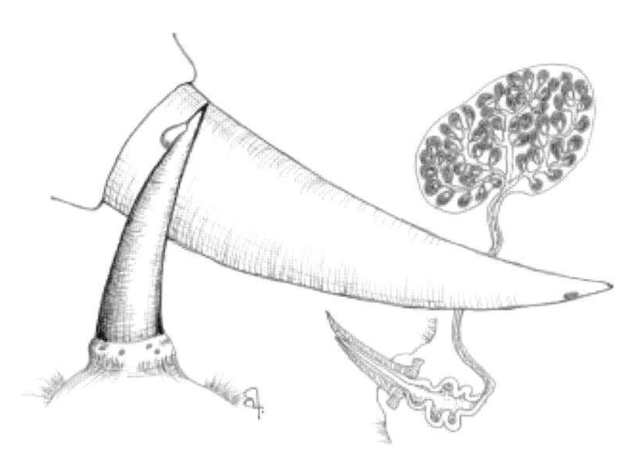

'I wounded one with small shot; and on my overseer's taking it out of the water it stuck its spurs into the palm and back of his right hand with such force, and retained them in with such strength, that they could not be withdrawn until it was killed. The hand instantly swelled to a prodigious bulk; and the inflammation having rapidly extended to his shoulder, he was in a few minutes threatened with locked-jaw, and exhibited all the symptoms of a person bitten by a venomous snake. The pain from the first was insupportable, and cold sweats and sickness of the stomach took place so alarmingly, that I found it necessary, besides the external application of oil and vinegar, to administer large quantities of the volatile alkali with opium, which I really think preserved his life. He was obliged to keep his bed for several days, and did not recover the perfect use of his hand for nine weeks.'

SIR JOHN JAMISON (1818)

In May 1991, while Keith Payne was fishing in the hinterland of Mackay in northern Queensland he noticed a platypus perched on a log beside the stream. When he approached more closely the animal did not move. Assuming it was probably injured or sick, he picked it up. Supporting it by one hand under the belly, he attempted to place it into the water. The animal suddenly wrapped its rear legs around his hand, driving one spur into the top of the hand and the other into the side of the ring finger. The following annotated excerpts from the medical report of the injury and its treatment relate a parallel series of events to those described in 1818 by Sir John Jamison:

> *There was immediate severe pain that was described as excruciating – much worse than previous shrapnel wounds the victim had suffered during the war [in Vietnam, where he was awarded the Victoria Cross for valour]. The pain and swelling became worse, reaching peak intensity about one and a half hours after the original envenomation. [Following a four and a half hour drive to a local general practitioner], intravenous injection of 15 mg of morphine was given slowly. Over the next five minutes very little pain relief occurred. Arriving at the hospital five and a half hours after the envenomation, he was still in severe pain and so shortly after admission a morphine infusion of 1.5 mg per hour was commenced, followed by a full right wrist block [anaesthetic].*
>
> *Forty-eight hours after the envenomation the hand and forearm were still tender to touch. Three days after, the swelling had decreased slightly and there was a little more movement in the fingers. Six days after admission the patient was discharged still with only 25% normal movement in his fingers. One month after envenomation the patient was still experiencing pain in his right hand and three months later he continued to have stiffness and swelling of the hand and fingers. At his last examination, 19 weeks after the incident, he still had pain, some swelling and reduced strength in fist-clenching movements.*

In December 2006 Payne reported still experiencing discomfort and stiffness in the injured hand when carrying out certain activities, such as using a hammer. He attributes the more restricted use of his right hand, at least partially, to the effects of being spurred by the platypus 15 years earlier.

Obviously, being spurred by a platypus is to be avoided if at all possible. Platypuses and people are seldom involved in any interaction that is close enough to result in the person being spurred. With the exception of anglers and individuals carrying out illegal netting activities, most people who come into such close contact with the species are researchers or zoo staff who know – or should know – that any platypus needs to be picked up only by the tail, just in case it is a male.

After a researcher was spurred when handling a platypus from the Duckmaloi River on the central tablelands of New South Wales, the authors of an article in the (British) *Journal of Hand Surgery* in 1994 suggested that 'warning signs should therefore be erected at air and sea ports warning tourists of the dangers of these venomous Australians'. Considering that just 17 cases of platypus envenomation in a century have been published and the Australian Venom Research Unit receives a request only every one or two years for advice in managing the effects of a human being spurred by a platypus, the suggestion of publicity for incoming travellers seems like something of an over-reaction!

Structure of the crural system

The pair of spurs in male platypuses are hollow and each is attached by a duct to a venom (crural) gland. The spur is made of keratin, sits on a small nub of bone (*os calcaris*) in the ankle region and is supported by muscles which can contract to evert it away from the ankle. The crural gland rests under the skin on the pelvis, surrounded by muscles of the buttocks and the body wall. From the gland, the duct conveying the venom passes between the muscles of the thigh to join the reservoir at the base of the spur (see Figure 3.1 and Colour Plate (top), page 35).

The short- and long-beaked echidnas are the only other living mammals that have spurs. In the echidnas the crural system does not appear to function in the delivery of venom. Juvenile male and female echidnas both have spurs but these are normally retained only in adult males. A spur and *os calcaris* have been found in the fossil remains from China of an

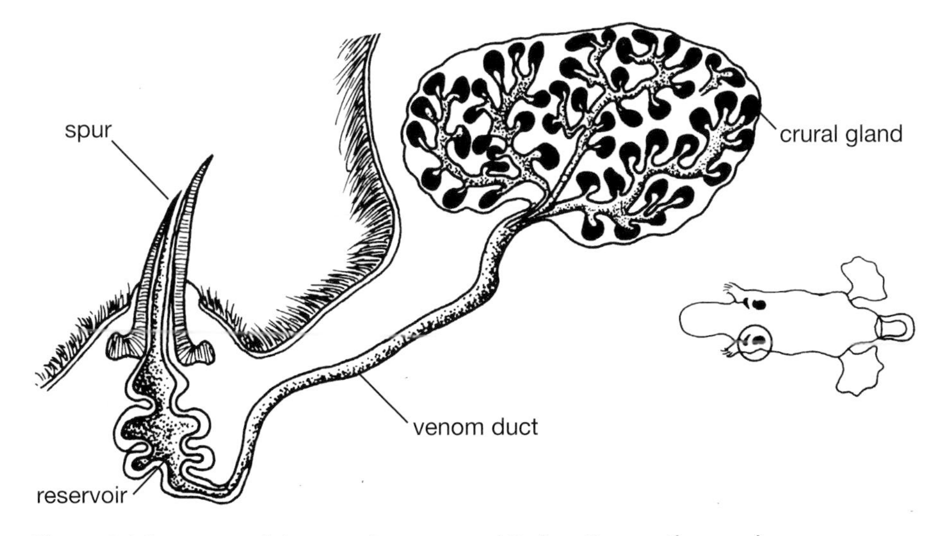

Figure 3.1 Structure of the crural system and its location on the rear legs.

ancient aquatic mammal, unrelated to the monotremes, and estimated to be around 164 million years old. *Os calcaris* bones have also been found in the rear limbs of a number of other fossils representing different groups of ancient mammals, and it has been suggested that perhaps many early mammals may have had spurs and venom glands.

In the male platypus a spur sheath grows in the nestling, inside which the spur develops and increases in size. The spur structure, size and development stages are shown in Figure 3.2, including measurements made from nestlings in museum collections. When the juvenile male emerges from the nesting burrow, the spur sheath – a chalky white conical structure – is present and persists for around six months after emergence (Stage 1). During the next three months the sheath breaks away to expose the spur itself, which is around 1.5 cm in length (Stage 2). Once the sheath breaks away completely, a pink collar of skin is evident, extending about one-third the height of the spur above its base. This prominent collar persists for another three to four months (Stage 3) but then recedes until it is just visible during the third year of the animal's life.

These morphological changes in the spur allow the age of young males to be determined. Animals with stages 1–3 can be recognised as juveniles from the most recent breeding season. Individuals with a collar that is quite evident but extends less than one-third of the way up the spur are in their second year of life, and those with a barely visible collar (Stage 4) are in at least their third year of life. After that time age cannot be determined, although some older animals may have spurs that have worn so that they are rounded, rather than pointed at the tips.

A spur does not develop in the female. However, a spur sheath of around 2 mm is present in the ankle region for 8–10 months after emergence in wild animals but is reported to be lost earlier in captive individuals. In certain wild animals the sheath on one hind limb may persist longer than that on the other. After the spur sheath is lost, a shallow, narrow depression in the skin on the inside of each ankle remains. A female with any remnant of a spur sheath is known to be a juvenile from the most recent breeding season. After the sheath is lost it is impossible to tell the age of a female platypus. Figure 3.3 illustrates the spur sheath structure, size and developmental stages in females, including nestlings from various museum collections, from hatching to the end of the first year of life, by which time the structure has been lost.

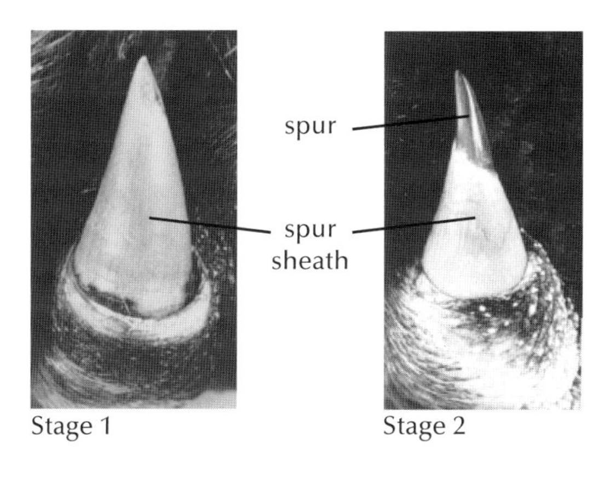

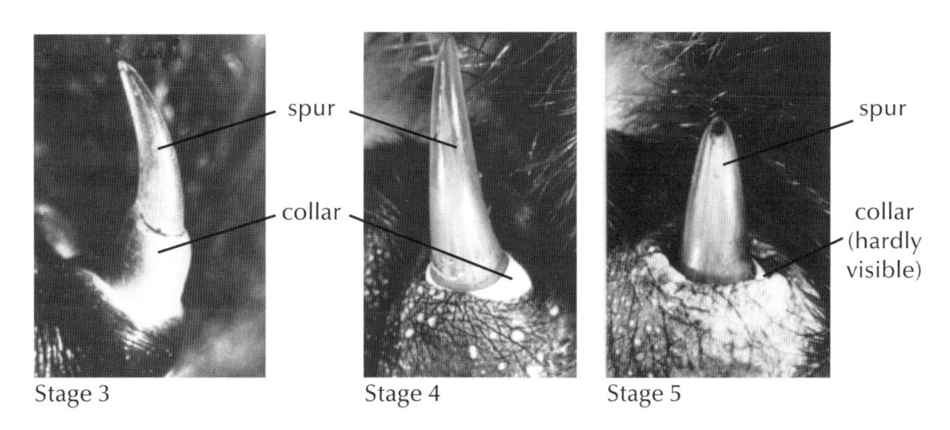

Figure 3.2a Spur development in the male platypus. Enclosed in the chalky sheath, the male spur grows then becomes visible as the sheath disintegrates. Stage 1: 0–6 months after emerging from the nesting burrow. Stage 2: 6–9 months from emerging. Stage 3: 9–12 months from emerging. Stage 4: more than 18 months from emerging. Stage 5: Old animal with tip worn from the spur. Photos: Peter Temple-Smith

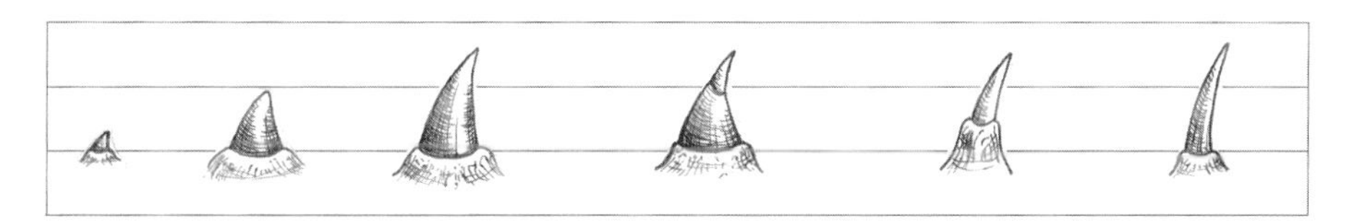

NESTLING				STAGE 1					STAGE 2				STAGE 3				ADULT		
OCT	NOV	DEC	JAN	FEB	MAR	APR	MAY	JUN	JUL	AUG	SEP	OCT	NOV	DEC	JAN	FEB	MAR	APR	MAY

Figure 3.2b Timing, development and size of the male 'spur' (spur and/or spur sheath) from nestling to adult. Horizontal lines in the diagram are 10 mm apart.

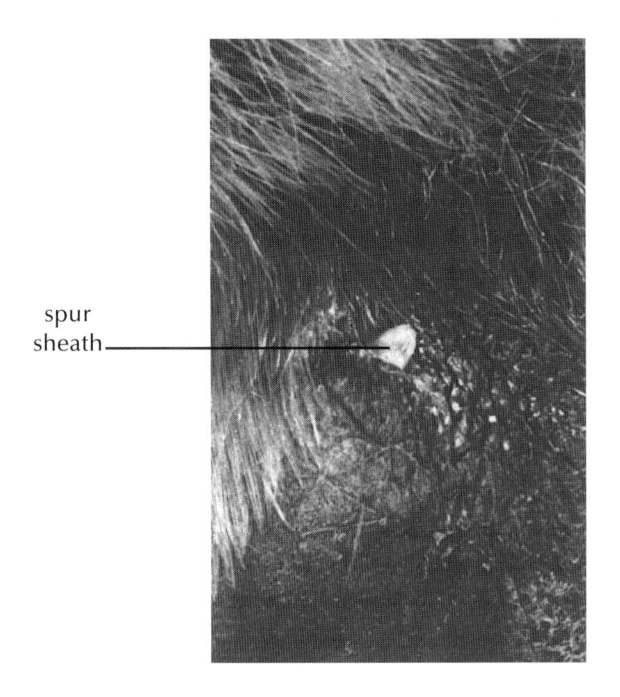

Figure 3.3a Remnant spur sheath on a newly emerged female which is retained only for a few months. Photo: Peter Temple-Smith

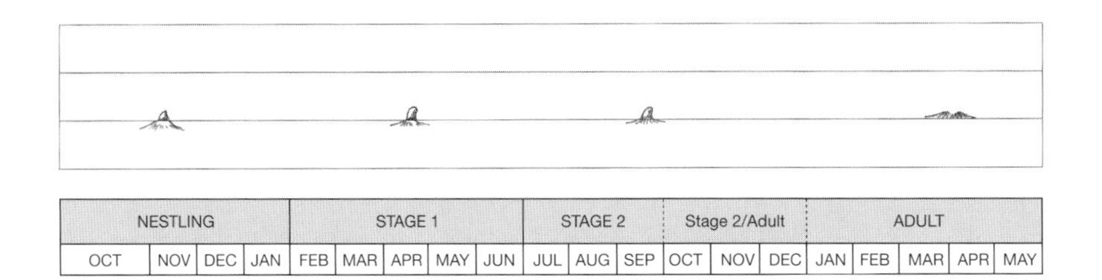

Figure 3.3b Timing, development and size of the female 'spur' (spur sheath) from nestling to adult. Horizontal lines in the diagram are 10 mm apart.

Venom

Especially during the breeding season when the crural glands are enlarged (Figure 2.2), venom sometimes is seen exuding as drops from the tips of the spurs of male platypuses during handling. It is a clear, slightly sticky liquid and consists of a complex mixture of molecules, including a range of proteins and peptides (chains of amino acids), many of which are unique to the platypus. Although there has been a good deal of biochemical investigation of the venom since as early as 1935, and more recently with the development of advanced methods of analysis, not all of its components have been identified and the functions of those which have been studied are not well understood.

Excruciating pain and long-lasting tenderness and sensitivity are the predominant symptoms attributed to platypus venom. The venom contains a nerve growth factor that is also found in snake venoms. This nerve growth factor is thought to contribute significantly to the pain experienced by victims and possibly has a direct effect on pain receptors. Molecules known as natriuretic peptides are found in the venom of the platypus and in other venomous species, such as the green mamba snake. The peptides cause dilation of blood vessels, reduction in blood pressure and the release of histamines resulting in inflammation and swelling. These compounds may also be involved in pain sensation and in the fall in blood pressure recorded in certain patients envenomated by the platypus. Four other interesting peptides have also been identified in the venom. In sea anemones and rattlesnakes, molecules similar to these peptides have been found to have neurotoxic (acting on nerve cells) and myotoxic (acting on muscle cells) effects, but neither whole platypus venom nor one of the peptides isolated from the venom had any effects on nerves or muscles. However, it has been suggested that they may also be involved in causing pain, as they are among the most abundant components of the platypus venom. Two of these four peptides have the same amino acid sequence, differing only because their second amino acid has an altered arrangement of its atoms in one (L-isomer) compared to the other (D-isomer). In addition, two of the natriuretic peptides differ in the same manner. The enzyme that produces such alterations, known as an L-to-D-peptide-isomerase, has recently been isolated from the platypus venom. This enzyme and the peptides resulting from its action have not been found previously in mammals.

Function of the crural system

The crural system (spurs and venom glands) may have been a more common feature in ancient mammals but why has it been retained as a potent venom delivery system only in the platypus, and why only in males and not females of the species? Although its function is a matter of debate, the possession of a venom with chemical characteristics conducive to causing severe pain has presumably given the modern platypus some evolutionary advantage.

It seems obvious that the venom delivery apparatus is not directed at predators, as presumably both sexes would have been exposed to the same predatory pressures throughout the five million years since this species (*Ornithorhynchus anatinus*) appeared in the Australian fossil record. The specialised structure of the bill leads to the presumption that the species has always relied on small macroinvertebrates as its predominant food source and it has never been required to subdue its prey. Retention of the system in males only, and the observation that the venom gland increases in size during the breeding season (Figure 2.2), suggest that the crural system is associated in some way with reproduction.

During the preparation for the Sydney Olympic Games in 2000, several Australian animals, including the platypus, were proposed as possible mascots for the games. A letter to the *Sydney Morning Herald* at the time advocated strongly against the platypus as a suitable candidate, because of the perceived violent behaviour of the male of the species towards the female during mating. This comment was motivated by the notion – disseminated by early naturalists and perpetuated in certain modern popular publications – that the spurs are used by the male to hold and subdue the female during coupling. In the wild, platypuses have been seen interacting in the water but their encounters are normally brief and difficult to observe, so that these observations do not support or refute the idea that females are treated this way during mating. The use of closed circuit monitoring, however, has permitted much closer examination of these interactions in captive animals. Observations of mating sequences from Taronga Zoo during the 1990/91 breeding season recorded bleeding from wounds on the tail of the female after two separate mating interactions. These wounds could have been inflicted by the spurs, but may also have been produced by the claws on the rear feet, which are quite sharp. The observations did not appear to indicate that the spurs were actively used to hold or subdue the female. The restraint of the female by the male using its spurs has not been documented by other studies of reproductive behaviour in captivity. Incidentally, the platypus (named 'Syd') did become one of the Sydney 2000 Olympic mascot species, although as a brightly coloured and somewhat stylised version of the real animal.

The other possible answer to the question of why the spur has been retained in the male platypus and not in the female is that males compete, either to maintain mating territories or access to mates, or both, and that this competition involves physical interaction, using the spurs to inflict painful but non-fatal injuries. This idea gains support from the increased size of the venom glands and the rise in aggressive behaviour of males during the breeding season. Males in captivity have been known to kill each other using their spurs, although capture of animals in the wild with fresh and healed spur wounds suggests that mortality does not normally occur. Presumably in the wild, animals have the opportunity to escape from encounters or to avoid them completely after an initial confrontation. In this way, interactions using spurs could be involved in the maintenance of a mating system in the species.

In Badger Creek and the Watts River in Victoria, some radio-tracked male platypuses appeared to have home ranges which were exclusive to themselves. Others had an overlap in their home ranges, but tended to avoid each other within the shared parts of their ranges. Also in Victoria, in the Goulburn River downstream from Eildon Weir, it was found that, while there was overlap of male platypus home ranges during both the breeding and non-breeding seasons, areas most used by the males became spatially separated from each other during the breeding season and certain males began to forage at different times of the day, suggesting temporal separation.

These observations, the seasonal changes in the crural system, and data from mark and recapture studies gave rise to the hypothesis that perhaps particular males in a platypus population are more successful than others and that they fertilise most of the breeding females each year. Under such a proposed mating system, it might be expected that a few males would make a disproportionately high genetic contribution to the population's offspring. Genetic studies carried out over a number of years in the upper Shoalhaven River in New South Wales in an attempt to test this hypothesis encountered a variety of obstacles and were never completed. However, the data from these studies did not appear to support the hypothesis. While the maternity of certain emerging young could be attributed to a number of female platypuses regularly captured in the area, the maternity of others was linked to females captured intermittently, or to unknown females, and the paternity of very few offspring could be attributed to those males captured more regularly in the area. Although incomplete, genetic studies of platypuses in the upper Shoalhaven River and of those introduced from Kangaroo Island to Warrawong Sanctuary in the Adelaide Hills tantalisingly suggested the possibility that at least some females may successfully mate with different males in the same breeding season. In each study a single pair of young from the same breeding season was reported to have had the same mother but different fathers.

The mating system of platypuses in the wild is obviously poorly understood. This is a significant gap in the understanding of the species' biology. The predictive and conservation values of environmental impact assessments, local recovery plans, translocations and other management strategies are weakened by such a lack of understanding. For example, it may be worthless conserving or rehabilitating a section of stream if that section is insufficiently large to support a critical number and/or mix of breeding, non-breeding and juvenile animals. Other aspects of the conservation biology of the platypus are discussed in Chapter 8.

4
THE SENSORY WORLD OF THE PLATYPUS

'My observations of the platypus under water support the view that, of the five senses ordinarily possessed by animals, the only one operative while the platypus is gathering its food at the bottom of a river or water-hole is that of touch. My opinion is that this animal must have developed some extraordinary means of finding its prey, apart from the sense of touch, and that the sensory apparatus through which it acts is connected in some way with the fleshy bill. If this 'sixth sense' is not responsible, then we must fall back upon that makeshift word 'instinct'.'

HARRY BURRELL (1927)

In the wild, platypuses can often be observed in the water continuing to forage, apparently unperturbed by a human presence. Yet the slightest sound or movement will cause them to disappear, presumably in response to something they hear or see. Most probably the species also tastes and smells food in its mouth. In other words, the platypus uses the familiar senses possessed by other mammal species. However, the nature of the sensory world experienced by a platypus embarking on its daily foraging can be described in the following way:

It was close to dusk when the platypus quietly slipped into the water. Momentarily floating under the overhanging shrubs, she could see the shapes of the cattle at the water's edge, silhouetted against the pink evening sky. The sound of the riffle upstream of the pool was quite loud and the water discoloured, as a result of recent welcome rains in the catchment. The scent of newly cut, but unbaled, hay floated in the air. As she moved out past the shelter of the bank, arched her back and dived, all of these sensations disappeared, her eyes, ears and nostrils instantly closing. Approaching the bottom she began to swing her bill gently from side to side, then touched the surface, disturbing sand and dislodging a few cobbles. Got it! A small freshwater crayfish held in her bill was broken up with the help of her front feet, the pieces were sorted into her cheek pouches and she headed back towards the surface.

The fact that a platypus closes its eyes, ears and nostrils when diving has resulted in a great deal of speculation over the years as to how it manages to seek out food along the river bottom. Platypuses housed in glass aquaria have occasionally been seen to open their eyes, leading to the proposition that perhaps vision is used under water. However, the species is mainly active at night, often also foraging in turbid waters, where vision is of little use. Most observers came to the conclusion that a special sense of touch was somehow involved, especially through the bill. Naturalist Harry Burrell observed that a burrowing platypus seemed able to excavate among rocks and roots of trees, somehow sensing their presence before actually striking them. He also constructed an elaborate tank where he could observe a captive platypus through a glass dome while inserting his hands into the tank through sealed openings. Although the platypus had been deliberately handled by him, in order to 'give her ample experience of the smell of human hands (to say nothing of the sight or feel of them)', the animal did not avoid approaching his hands in the tank. From this single experiment he concluded that the platypus was indeed not using its senses of smell, vision or hearing when under water. The species was obviously able to find its food without using these senses and he concluded that, although the sense of touch could be involved, another sense was present in the platypus – in fact, a 'sixth sense'.

Vision

The eyes of the platypus are quite small (6 mm), spherical, have a round pupil and are enclosed in grooves which also house the ear opening (see Colour Plates, pages 33 and 40). No investigation has been done but, because of the placement of the eyes relatively high and on the sides of the head, it seems unlikely that the animal has much in the way of stereoscopic (binocular) vision and the associated visual depth perception. The platypus has no eyelid but, like birds and reptiles, it has a nictitating membrane, which moves over the eye to clean and lubricate it by dispersing secretions from the lacrimal (tear) glands. Investigation of an eye preserved in whisky in 1884 showed that the retina has both rods (sensitive to light intensity) and cones (sensitive to colour) but the sensitivity of the eye to light intensity or the extent of its colour vision have not been determined. The area of the brain that receives nerve input from the eyes is relatively small. However, observations of the platypus in the wild suggest that it has acute eyesight.

Interestingly, the back of the lens is considerably more convex than the front, and the cornea is relatively flat. These characteristics and the position of the area of greatest sensitivity (*area centralis*) are all very similar to those found in mammals that have good underwater vision (e.g. otters and seals), suggesting perhaps that the ancestors of the platypus may have used vision to hunt larger prey items in water (Chapter 7).

Hearing

Sound waves enter the ear via quite a long (around 4 cm) outer ear canal, that extends from the surface to the tympanic membrane (ear drum). As in all mammals, the sound is then transmitted and amplified through the middle-ear to the oval window of the inner-ear by the movements of the three middle-ear bones (malleus, incus and stapes). Middle-ear bones found in a species of fossil platypus (*Teinolophos trusleri*), which is estimated to be 110 million years old, suggest that the evolutionary change resulting in these three bones being involved in sound transmission may have occurred independently in the ancestors of monotremes and those of the therian (marsupial and eutherian) mammals. As the ear openings are closed during diving, it appears likely that hearing may not be used in that medium. However, sound is transmitted particularly well in water. It has been suggested that sound waves striking the side of the head may resonate against air trapped in the canal of the outer ear, permitting certain sound frequencies to be heard under water.

The cochlea of the inner-ear in the platypus is predominantly mammalian in its appearance but is less coiled than in other mammals and has sensory cells arranged a bit like those found in the ears of birds. The cochlea is sensitive to airborne sounds of frequencies between 500 Hz (Hertz) and 20 kHz (kilo Hertz). These frequencies are much the same as those detected by the human ear. In the platypus, the cochlea is most sensitive to lower frequencies (lower pitched sound) of around 4 kHz. Like many other animals, the platypus probably also detects sounds from the environment that are conducted to the inner-ear through the bones of the skull. The area of the platypus brain that receives information via the nerves from the ears is relatively small and overlaps the visual area.

Smell and taste

In the short-beaked echidna the olfactory bulb at the front of the brain is more developed than in the platypus (Figure 4.1). Interestingly, recent studies of the skull of the fossil platypus *Obdurodon dicksoni* (Chapter 7), using high resolution X-ray computed tomography scans also indicated a small olfactory bulb in that species. The nasal passages of the platypus are lined with sensory cells but there are fewer fine nasal bones (turbinals) covered with sensory cells than in the short-beaked echidna, which almost

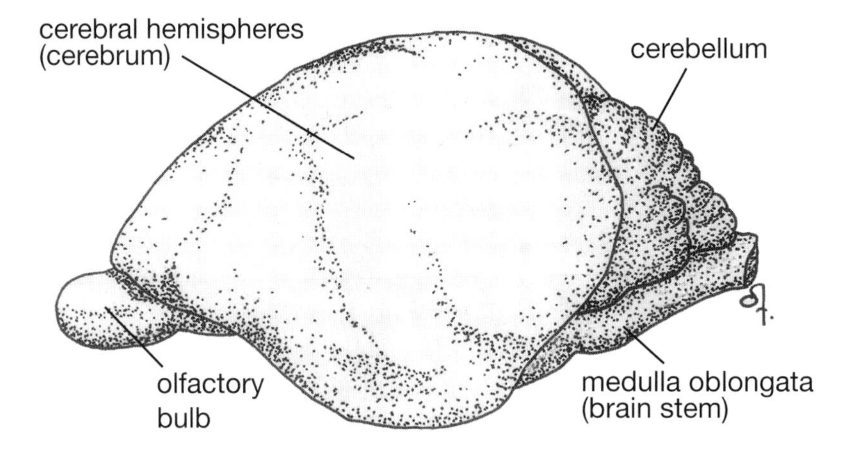

Figure 4.1 Side view of a platypus brain. The cerebrum is involved with the senses and conscious activities, the cerebellum with coordination, the brain stem with involuntary control (e.g. heart beat and breathing) and the olfactory bulb with smell.

certainly uses its sense of smell (olfaction) in seeking out its food – ants and termites – and in the location of females by males during the breeding season. The nostrils of the platypus are closed by valves when underwater, so that olfaction is probably not involved in foraging behaviour. The platypus has two well-developed and functional scent glands located in the shoulder/neck (cervical) region. These appear to increase in activity during the breeding season, giving a 'musky' odour, particularly to the males. There have been reports of platypuses apparently 'marking' objects, like rocks or logs, beside or under the water, by rubbing against them. These observations, and the possession of specialised scent glands, seem to suggest an important, but unknown, function of the sense of smell.

At the back of the tongue, the platypus has two grooves lined with sensory cells (circumvallate papillae), which are probably involved in tasting food in the mouth. There is also a Jacobson's organ consisting of two pouches in the roof of the front part of the mouth, which may also be involved in the sense of taste/smell of food in the mouth.

Special touch: the 'sixth sense'?

From as early as 1885, anatomical investigations of the bill had revealed specialised nerve endings associated with pores on its surface (Figure 4.2). The early scientists also noticed extensive nerve connections in the bill and the large size of the trigeminal nerve linking the bill to the brain. They correctly attributed a mechanoreceptor function to the pore-linked structures; i.e. they were involved somehow in the reception of touch (tactile) stimuli. However, they also found a type of small glandular structure, again opening via pores on the surface of the bill. These two distinct structures were later identified and described in detail by sophisticated light and electron microscopy.

The first type, the rod organs or push rods (the mechanoreceptors), occur in large numbers (46 500–60 000) and are found regularly distributed over the bill, but are more numerous on the upper than on the lower bill and are most densely congregated along the border of the upper bill. The other type, the glandular structures, are less numerous (around 40 000) and can be categorised into three sub-types, only two of which have nerve connections. The most numerous of these open into the largest pores on the bill and are called sensory mucous glands.

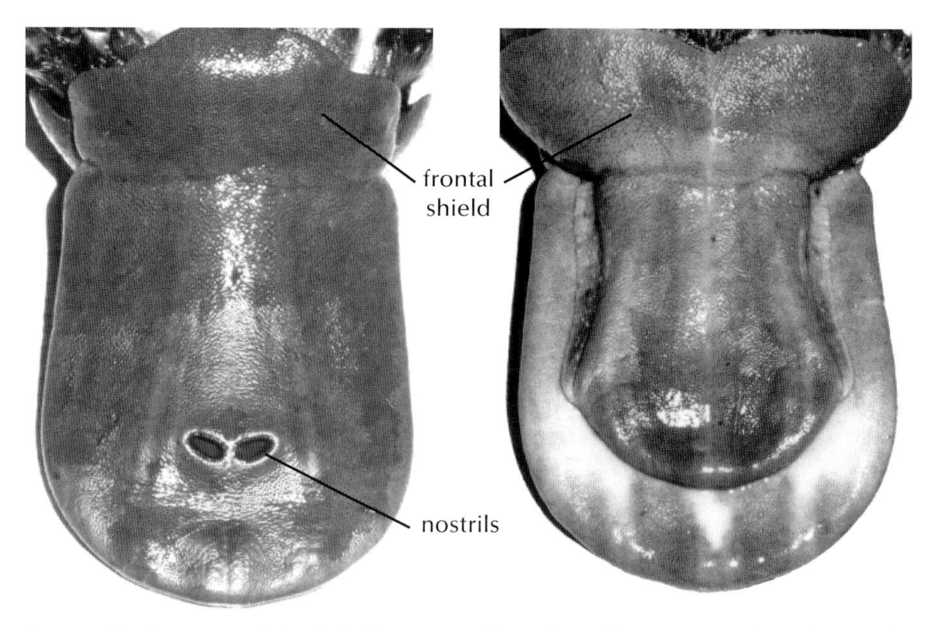

Figure 4.2a Structure of the bill. The upper (dorsal) and lower (ventral) surfaces of the bill are shown. Photos: Ros Bohringer

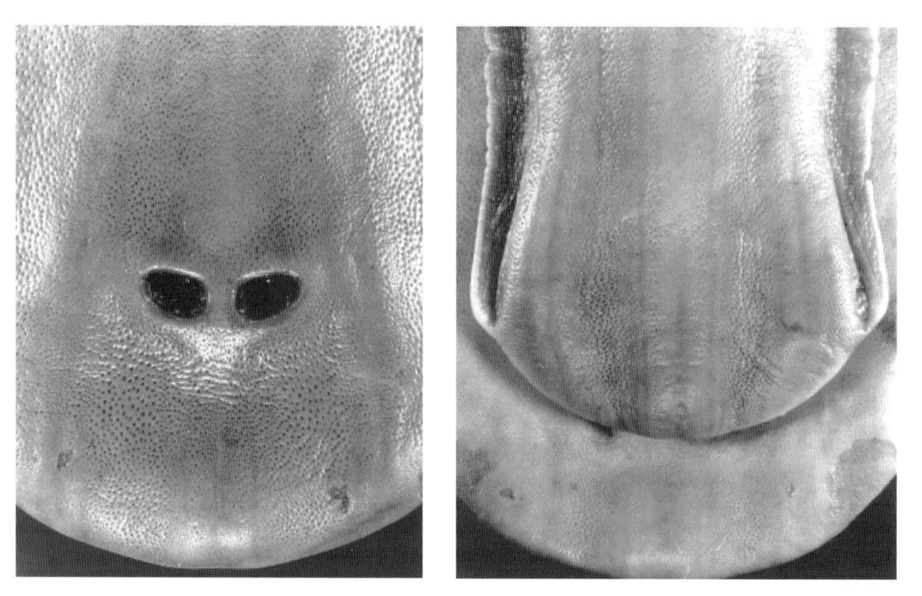

Figure 4.2b Enlargement of upper and lower bills showing sensory pores. More are present on the dorsal (left) than on the ventral (right) bill. Photos: Uwe Proske

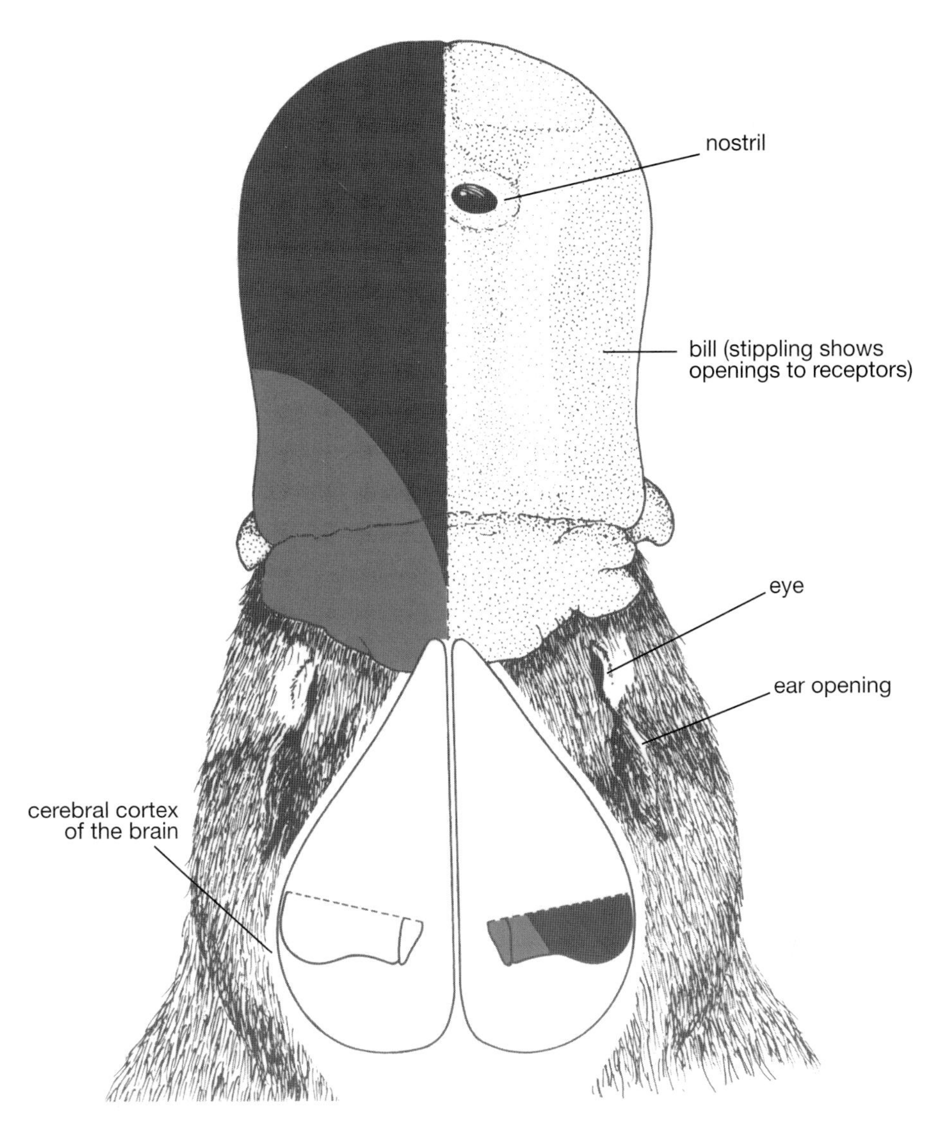

Figure 4.3 The skin receptors in various parts of the bill send impulses to regions on the opposite side of the cerebral cortex of the brain (half bill figures shown on the brain) from the side of the bill actually receiving the stimuli. Diagram: Uwe Proske

In most species of mammals the cerebral hemispheres (or cerebrum) in the brain are noticeably folded. It has been suggested that the amount of folding in a species is an indication of its evolutionary progression. However, this idea is not supported in the monotremes, where the platypus has cerebral hemispheres that show practically no folding (Figure 4.1), while those of its closest relative, the short-beaked echidna are distinctly folded. The cerebrum is the part of the brain involved in receiving nervous input from sense organs (eye, ear, nasal passages, taste buds and areas of touch sensation), which provide information about an animal's external environment. The most superficial layer of the cerebrum is the cerebral cortex, the posterior (back) part of which receives information from the sense organs. The anterior (front) part processes this sensory information and the middle area (motor cortex) mediates the resultant functions of the body (e.g. feeding activity, nesting behaviour, avoidance of predators). During the mid-1970s, sophisticated recordings were made of nerve input to the sensory cortex of the brains of anaesthetised platypuses, when stimuli were applied to various parts of the body. These recordings permitted the areas receiving sensory information to be mapped. Perhaps not surprisingly, these studies indicated that the bill had by far the largest area of sensory reception in the cortex, with the auditory and visual areas being much smaller. In Figure 4.3 the areas of the cerebral cortex receiving information from the bill are shown. Nerve impulses from receptors on the right side of the bill are received on the left side of the brain and vice versa.

The huge number of push rod mechanoreceptors and their distribution on the bill led to the proposition that the function of the bill in the location of food was almost certainly a tactile one. The hypothesis was that water movements that emanate from a moving prey item stimulate the push rod receptors, enabling the prey to be detected close to the bill, perhaps as far away as about 10 cm, but without the bill actually needing to touch it physically. Also, it was considered possible that tactile stimuli were received through the movement of water around objects in the underwater environment, generated by the movement of the stream and/or by the platypus itself. Perhaps this was the 'sixth sense' which permitted the diving platypus, essentially deaf and blind, to locate its food and navigate underwater. Certainly, the presence of huge numbers of push rods spread over the bill, and their extensive nervous representation in the sensory cortex of the brain, suggested their importance in sensory perception.

Electroreception: the 'sixth sense' revisited

Several species of fish, including certain sharks and rays, have receptors in pores that are known to detect prey animals by picking up the small electric fields generated by their muscle contractions. Such a sense had not been reported in mammals until a German neurobiologist, who had carried out research on electroreceptors in fish, noticed the similarity between the sensory mucous glands of the platypus and the structures detecting electric fields in these fishes. A simple initial experiment with captive platypuses showed the apparent detection of a small 1.5 volt battery in a holding tank. Recordings from the sensory cortex of two anaesthetised animals subsequently showed that this area of the brain was receiving nervous input produced by electrical stimuli to the bill. In fish species where electroreception occurs, the electrosense is related to the lateral line system, which detects small pressure changes in the water generated by moving objects. In these species, this information reaches the brain by way of the vagus nerve (10th cranial nerve), while in the platypus, nerves from the electroreceptors and mechanoreceptors in the bill are connected to the trigeminal nerve (5th cranial nerve), which is very well developed in the modern platypus and, according to recent studies, was probably also prominent in the fossil platypus, *Obdurodon dicksoni* (Chapter 7).

Electroreception is known to have evolved independently at least twice in fishes. It is interesting that a sophisticated electrosense has also evolved in the platypus, but has arisen from completely different parts of the nervous system and is also associated closely with the tactile system in the bill. These important experiments and observations were reported in the prestigious scientific journal *Nature* at the beginning of 1986, and have assumed something akin to the historical significance of William Caldwell's 1884 announcement that the monotremes laid eggs.

During the 1990s a great deal of sophisticated research was carried out, including a demonstration of the presence of electroreceptors in both the short- and long-beaked echidnas. In the echidnas there are fewer of these receptors than in the platypus and they are more restricted in their distribution on the muzzle (beak). Mapping of the parts of the sensory cortex receiving information from the electroreceptors in the bill of the platypus showed that these areas almost completely overlapped and extended beyond the extensive region receiving input from mechanoreceptors (tactile stimuli). It appeared that electroreception was, after all, the 'sixth sense' proposed by Harry Burrell. However, as often happens in science, more knowledge led to more unanswered questions. The electric fields apparently necessary to generate impulses relayed from the bill to the sensory cortex were 1–2 millivolts per centimetre (mV per cm), while those generated by some of its prey were measured in microvolts (μV per cm). For example, a caddis fly larva, belonging to a group of insects which constitutes the predominant food source of platypuses in many areas (Chapter 5), generated a field of only 14 μV per cm. Larger prey items, like dragon fly nymphs and freshwater crayfishes ('yabbies'), produced fields of 150–580 μV per cm. Only one of the species studied, a common freshwater shrimp, was found to generate an electric field (1.9 mV per cm) within the range known to be detectable by the platypus.

So, is it back to the drawing board? Not necessarily. Electric fields generated by prey species are very difficult to measure accurately and measurements from the brain were made on anaesthetised animals. Although a few scientists still remain unconvinced that there is in fact a functional electrosense in the platypus, many further experiments with nerves, brains and close observations of the animal's behaviour in captivity have led to hypotheses and speculations relating to how an electrosense may function in a platypus that is actively foraging in the wild. The current hypothesis – as yet experimentally untested but supported by a range of data and theoretical considerations – suggests that the summation of the small electric signals reaching very large numbers of electroreceptors, essentially at the same time, would be sufficient to produce the required threshold for detection by the sensory cortex in the brain. In other words, it is the whole bill being stimulated which relays the necessary information to the brain. The available data from various experiments, and knowledge of the fine anatomical structure of the bill, have also resulted in the generation of hypotheses as to how the platypus senses the presence of a food item and determines its exact location in terms of both direction and distance.

It has been proposed that, as the platypus moves its bill from side to side while foraging, the polarity (positive and negative) and strength of the stimuli from the prey changes as the bill moves across the electric field. The electroreceptors are not evenly distributed over the bill, but are arranged in parallel rows running from its tip to the frontal shield. This results in discontinuous bursts of impulses being relayed to the brain as each row of receptors in the bill is swept across the electric field, possibly permitting the platypus to assess the direction of the source of the field.

Electric fields are propagated through water at a faster rate than are water movements. A moving prey item would generate both electrical (from its contracting muscles) and tactile (from its movement in the water) stimuli which would be received by the bill. Electroreceptors (sensory mucous glands) would be stimulated first, followed by the mechanoreceptors (push rods). Processing of this small difference in the timing of reception could permit the brain to assess the distance to the prey item. Even those who have proposed these possible ways in which the 'sixth sense' might work in the detection of direction and distance admit that their ideas are speculative and are yet to be formalised and tested experimentally. However, these are still fascinating proposals that attempt to explain how the platypus finds very small prey items in the middle of the night and in turbid water with its eyes, ears and nostrils closed. Other aspects of feeding are discussed in detail in the following chapter.

5
ENERGETICS, DIVING AND FORAGING

'In the dusk of the evening, I took a stroll along a chain of ponds (which in the dry country represents the course of a river) & had the good fortune to see several of the famous Platypus or Ornithorhynchus paradoxus. They were diving & playing in the water; but very little of their bodies were visible ...'

CHARLES DARWIN'S DIARY ENTRY (19 JANUARY 1836)

On a summer evening in 1836, Charles Darwin watched several platypuses foraging in a section of the Cox's River now impounded in the waters of Lake Wallace in New South Wales. He noted the animal's low profile in the water but also mentioned its diving habit. Similar observations of platypus diving and foraging behaviour were made by other Europeans – often hunters who commented on how difficult it was to take accurate aim due to the short time the animal spent on the surface of the water between dives. Because of its dependence for food on small invertebrate animals occurring at the bottom of the bodies of water in which it is found, the platypus spends a great deal of its time each day involved in the foraging and diving activities noted by these early observers and hunters.

Foraging

Foraging is mainly nocturnal but some individuals within a population, at certain times of the year and in some particular areas, appear to feed during daylight hours. Diurnal activity, however, seems to be the exception rather than the rule. The reason platypuses are normally seen during the early and later parts of the day is because they tend to leave their burrows an hour or two before it gets dark and return over a similar period after first light. Foraging may not be continuous and some animals leave the water to occupy one or more burrows for varying lengths of time during a normal feeding session. Average foraging periods last for around 10–12 hours per day but, in order to gather the small organisms on which it feeds, the platypus may sometimes forage continuously for 24 hours or more. Activity periods and distances individuals move during foraging have been found to vary considerably from one location to another and even between individuals from the same location.

Although individual platypuses may forage overnight in a single pool or short section of stream, others can traverse several kilometres during a 24-hour period. During their feeding activity, which includes diving as well as swimming on the surface, platypuses have been found to swim at speeds of 0.7 to 2.4 km per hour. In a study carried out at the University of Tasmania, the energy demands of platypuses swimming against different water flows through a tank showed the lowest energetic cost of swimming was at a speed of 0.4 metres per second (1.44 km per hour). Platypuses have been recorded swimming against currents of slightly over one metre per second (3.6 km per hour) but to do this they must exert a good deal more energy than they would during foraging in still or slowly moving water. The platypus has a lower body temperature than most other mammals. Consequently, the energetic cost of temperature maintenance is less than that for other mammal species. But the energy expenditure associated with entering cold water to feed during winter is considerable and the costs of swimming are also high. The buoyancy created by air trapped in the fur increases the energy demand of diving and more energy is expended in locomotion using the front feet (pectoral rowing) than in kicking with the back feet, as occurs in most other small semi-aquatic mammals.

Studies have shown that platypuses feed in both the slow-moving and rapid (riffle) reaches of streams, and that they often prefer coarser bottom substrates, particularly cobbles and gravel. The proportional presence of different substrate types in the lower Hastings River in New South Wales and the percentage of dives recorded on these substrates are shown in Figure 5.1. All types of substrates were used for foraging but few dives were seen over muddy substrates. While cobbled substrates made up only 12.7% of the study area, 28.6% of the recorded foraging dives by platypuses were over this substrate type, indicating a preferential selection. Although gravel made up over half of the available substrate, only 26.8% of the foraging dives occurred in gravel sections of the stream. This observation was unexpected, as foraging on gravel surfaces has been reported in other studies. The area of river bottom available to individuals for feeding presumably determines the number of platypuses a stream is capable of supporting (its carrying capacity). The capacity of streams to provide food for non-breeding animals, and for a proportion of lactating female platypuses during the breeding season, has not yet been quantitatively studied. Certain rivers obviously will be more productive than others and therefore support higher populations of platypuses. However, the conservation of platypus populations in any stream depends on the maintenance of its productivity, on which a variety of human-induced changes can have an impact, including the effects of disturbing riparian vegetation, erosion and changes in stream flows. These issues are discussed further in Chapter 8.

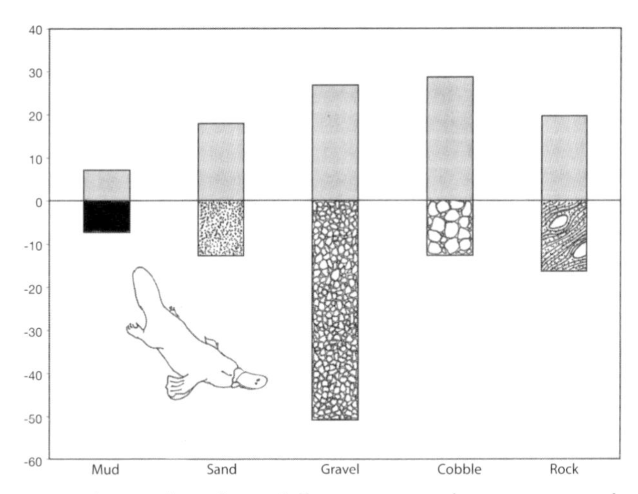

Figure 5.1 Platypuses foraged preferentially on some substrates more than on others in the lower Hastings River in New South Wales. The chart shows the percentage number of platypuses (stippled columns) observed diving over mud, sand, gravel, cobble and rocky substrates. (Negative % represent the proportions of each substrate occurring in the study area.)

Diving

A host of small invertebrate animals are found on the bottoms of rivers, creeks and other bodies of fresh water. Some of these are microscopic, but many are large enough to be seen with the naked eye. They include crustaceans, worms and molluscs (snails and freshwater mussels), along with adult and larval stages of many insects (e.g. water beetles, water bugs, dragonflies, caddisflies and mayflies). These are the macroinvertebrates, which are the predominant food source of the platypus. Although these animals can be seen, most are still quite small and the platypus must dive very frequently to find and use them as a food resource. For example, in Lake Lea, a sub-alpine lake in Tasmania, radio-tracked platypuses were found diving up to 1600 times during a single feeding session and on average plunging to the bottom of the lake 75 times every hour.

Diving times, lasting between 30 to 140 seconds, have been reported by both early naturalists and modern field biologists, with shorter periods of around 10–15 seconds spent on the surface between dives. In addition to this regular diving sequence during normal foraging, captive platypuses can be observed 'wedging' themselves under submerged objects for up to 11 minutes. A platypus achieves this by remaining very still, buoyancy holds the animal in place under the submerged object and it does not have to use up oxygen to produce energy for exercising muscles normally used while swimming underwater. Whether platypuses in the wild do this is not known. However, also in the Lake Lea area, biologists have observed platypuses 'wedging' themselves under ice and approaching a breathing hole only occasionally to obtain air.

The platypus relies on the oxygen content of the air in its lungs during diving. Like many diving birds and small mammals, it simply holds its breath to dive. However, the species does show a reduction in metabolic activity during diving, with a marked diving bradycardia, where the normal heart rate of 140–230 beats per minute drops to 10–120 beats per minute. An acceleration of the heart rate (tachycardia) to 300–400 beats per minute was recorded immediately before and after dives in one study. When presumed to be sleeping in its nest box, the heart rate of this platypus ranged between 75 and 120 beats per minute, with an average rate of 99 beats per minute. The platypus also has the ability to quickly inhale a large quantity of air into its relatively large lungs prior to descending. During its normal diving, metabolism apparently remains aerobic, probably only becoming anaerobic following dives lasting longer than the calculated aerobic dive limit of 40–59 seconds. In the Lake Lea study, 72% of monitored dives were 18–40 seconds in length, with slight increases in 'recovery' time at the surface after extended dives. Compared to that of other mammal species that do not dive or live in burrows, platypus blood has a high oxygen-carrying capacity due to the high haematocrit (49% of the blood is made up of red cells), large numbers of very small red cells (10 million cells per cubic mL of blood) and elevated haemoglobin levels (19 grams per 100 mL of blood). Other small diving mammals (e.g. musk rats) or ones that occupy burrows where oxygen levels are low (e.g. ground squirrels) show similar haematological characteristics to those of the platypus.

In January 1996, biologists from the New South Wales Fisheries found a drowned platypus in a fishing net set at a depth of about 30 metres in Lake Yarrunga, a water storage impoundment behind Tallowa Dam at the junction of the Shoalhaven and Kangaroo Rivers in New South Wales. This was unusual, as platypuses are seldom found in such deep lakes, presumably because of their limited diving abilities and the energetic costs of having to dive to such great depths to collect quite small food items. Platypuses normally forage in pools of up to 5 metres in depth. In Lake Lea, for example, the average depth of monitored dives was 1.3 metres and 98% of dives were to depths of less than 3 metres. The maximum diving depth in that study was 8.8 metres. Figure 5.2 shows the percentage

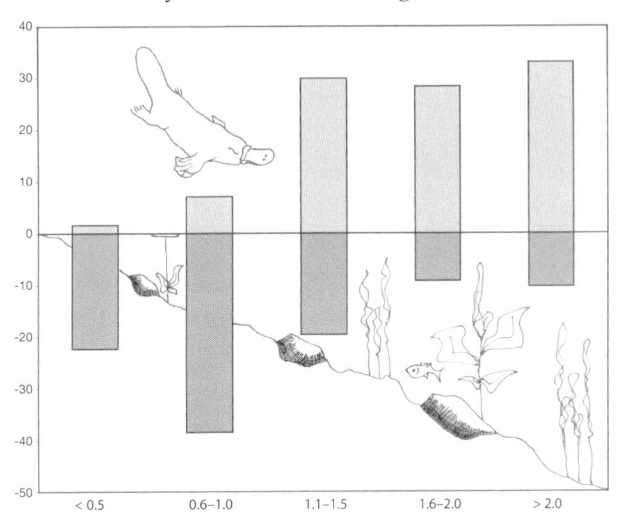

Figure 5.2 Platypuses use the deeper areas of the Hastings River in New South Wales. Percentage number of platypuses (stippled columns) observed diving to depths of less than 0.5, 0.6–1.0, 1.1–1.5, 1.6–2.0 and deeper than 2.0 metres (negative % represent the proportions of each depth category occurring in the study area) are shown.

representation of various depth categories in a study area in the lower Hastings River in New South Wales, and the percentage of platypuses diving in each depth category. Animals seemed to prefer foraging in water deeper than one metre but it is not known if the recorded foraging depths in these studies are related to the availability of food organisms at those depths or to another factor, such as avoidance of predation by foxes or even birds of prey (Chapter 8).

Diet

The platypus has a relatively simple digestive system. Both its small and large intestines are quite short and its stomach extremely small. The 'milk' teeth of the young platypus are replaced early in its life by horny grinding pads made of keratin that continue to be replaced throughout life. These grind the food particles very finely, aided by sand and other inanimate material taken in while feeding. As a result, the contents of the digestive system and material remaining in the droppings are so small that they are all but useless for determining what the platypus has been eating. As well, the faeces are black, semi-solid and very foul-smelling indeed. Certain platypuses, particularly in Tasmania, will deposit these on land but most pass their faeces into the water where they cannot be collected for analysis. While foraging, the platypus stores food in its cheek pouches which lie beside the horny grinding pads in the mouth. Information regarding the food consumed by the species has been studied since the early 1800s by investigating the contents of these cheek pouches. Food items are stored in these pouches during foraging on the river bottom before being masticated when the platypus rests on the surface. Characteristic ripples can be seen moving out from around the bill as the grinding process takes place. Indigestible particles are thought to be separated out by movement through the distinctive soft grooves in the lower bill. The early naturalists emptied the cheek pouches of animals which had been killed, but biologists now use a small stainless steel spatula to remove material from the cheek pouches of living platypuses captured in the wild (Figure 5.3).

Figures 5.4a and 5.4b show a range of prey species and the nature of the fragmented material used to identify food items under the microscope. While considerable numbers of unidentified particles do occur, the presence of benthic macroinvertebrate species, especially insect larvae, have been found by most investigators in the cheek pouches. Free-swimming

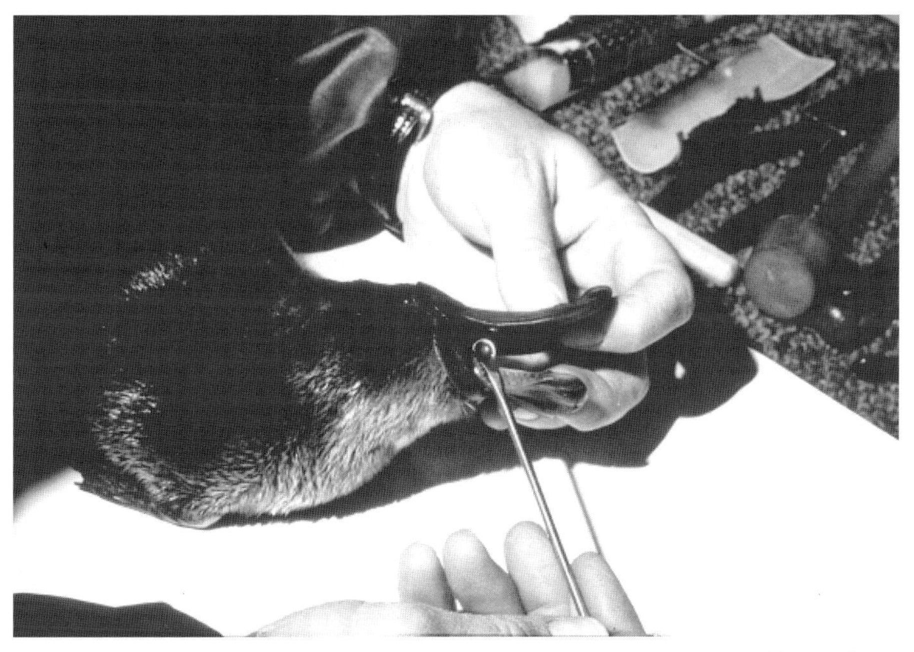

Figure 5.3 The diet of the platypus has been investigated by using a small spatula to remove ground-up food material from the cheek pouches of captured animals.

species, such as shrimps, swimming beetles, water boatmen and backswimmers (water bugs) and tadpoles, also appear to be occasionally included in the diet along with worms, small freshwater pea mussels and snails. There have been occasional reports of cicadas and moths being taken from the water surface as well. Platypuses are sometimes caught on fishing lines, having obviously been attracted to live bait, spinners and even artificial flies.

Platypuses readily feed on freshwater crayfish (yabbies, usually *Cherax destructor*) of up to around 10–15 cm in size provided to them in captivity. Cheek pouch samples from Lake Lea in Tasmania were found to contain remnants of the small indigenous burrowing crayfish, *Parastacoides tasmanicus*, and this species was considered to constitute an important part of the diet of the resident platypuses. Although a number of crayfish species are found in various water bodies occupied by the platypus, few remains of any species of freshwater crayfish have been reported from cheek pouch material collected in other studies. There are a number of possible reasons for the disparity between the captive situation and the various studies. Because the crayfish are less numerous, and probably more patchily

Figure 5.4a Platypus food: an assortment of macroinvertebrate animals collected from the bottom of the Wingecarribee River in New South Wales. Photo: John Matthews

Figure 5.4b Ground-up food particles taken from the cheek pouch of a platypus captured in the Wingecarribee River in New South Wales. Photo: John Matthews

distributed than the other food items, the lack of crayfish remnants found in cheek pouches may simply reflect the fact that they are located only occasionally. When they are eaten, they presumably represent an excellent catch per unit effort for an individual platypus. It is also possible that, because the crayfish are larger than most other macroinvertebrate prey species, they may be processed by the platypus in a different manner, not necessarily resulting in body fragments accumulating in the cheek pouches. Certainly in captivity, a platypus will often grasp a yabby in its bill and rip it apart with the aid of the front feet, snapping at the various pieces released as a result of this activity.

The investigation of cheek pouch contents may not comprehensively identify all items that make up the diet of the platypus in certain areas. However, most field investigations to date indicate that the species appears to be largely dependent on bottom-dwelling (benthic) macroinvertebrates for food and is normally a dietary opportunist, largely exploiting the temporal and spatial availability of these macroinvertebrates. Figure 5.5 shows the percentages of various macroinvertebrates taken in summer and winter in the upper Shoalhaven River in New South Wales. Percentages of each type of prey varied, but caddisfly larvae constituted 41% and 64% of the diet in winter and summer respectively. This ubiquitous group of macroinvertebrates has been found to be the dominant group of prey species in several other areas as well. The freshwater shrimp (*Paratya australiensis*) was present year-round in the upper Shoalhaven River study area but was taken by the platypuses only in the winter, when the shrimps were found on the bottom of the stream. During other seasons these shrimps were found swimming in mid-water depths and were taken by the introduced brown trout (*Salmo trutta*) but not by the platypus.

In 1817 Sir John Jamison reported finding nothing else but 'small fish and fry' in platypus stomachs. Morton Allport, in 1878, was so concerned about platypuses eating trout fingerlings that he stated, 'I cannot conscientiously recommend the owners of trout streams to encourage the presence of the *Ornithorhynchus anatinus*', and as recently as 1958 an article in the *Australian Journal of Marine and Freshwater Research* reported platypus predation of trout fingerlings in hatcheries in Tasmania. In this latter instance, captured platypuses apparently regurgitated the partly eaten trout fingerlings of up to three inches in length and were reported as only consuming the heads and leaving the tail. In captivity, a number of individuals caught and ate plague minnows (*Gambusia holbrooki*) introduced

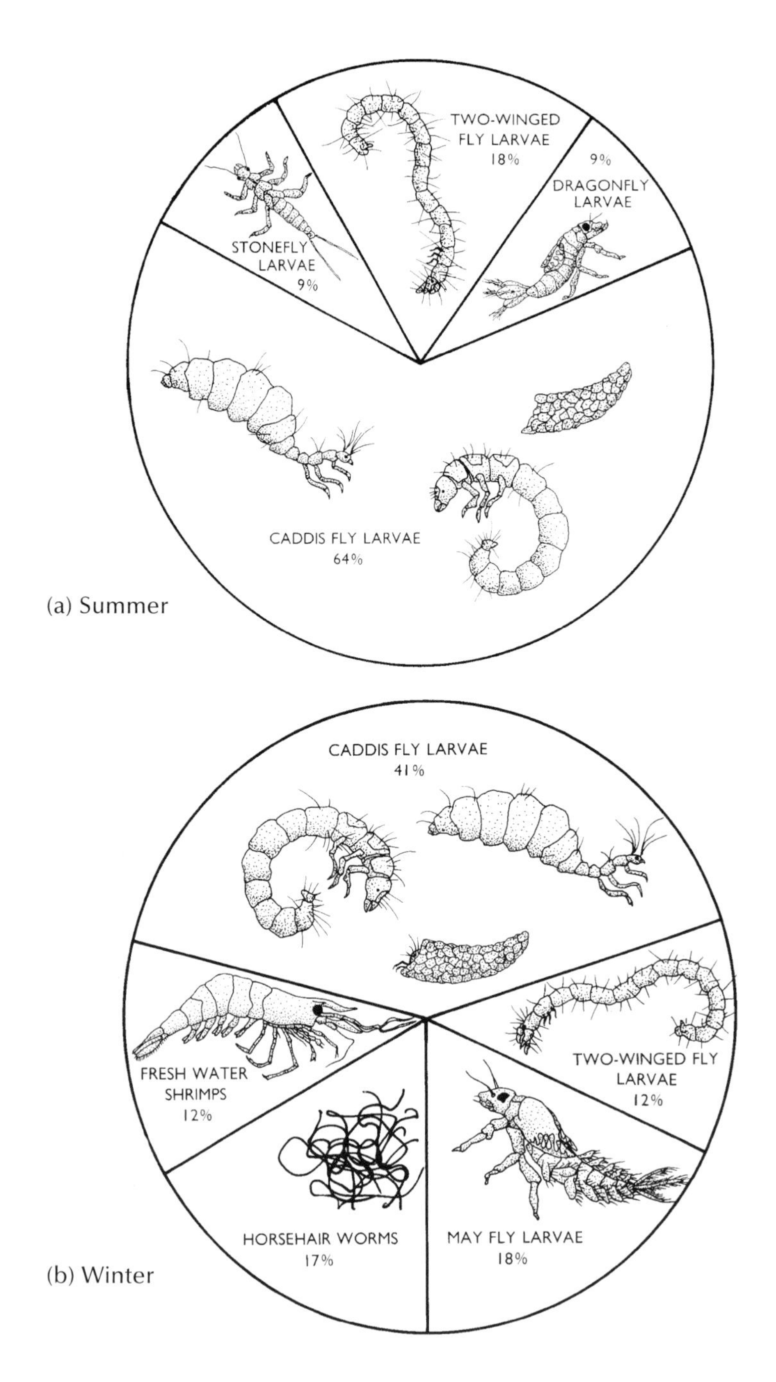

Figure 5.5 Macroinvertebrate food found in cheek pouch samples from platypuses in the upper Shoalhaven River in New South Wales in summer (a) and winter (b). The percentages varied slightly between seasons.

into the tank. Tiny calcium carbonate structures (otoliths) from the inner-ears of small fishes, presumably trout, were found in cheek pouches from platypuses in Lake Lea in Tasmania, but other recent studies of wild-caught platypuses have yielded no material (e.g. scales, bones or otoliths) attributable to fish species.

Introduced trout species show overlap in their diets with the platypus. Both the rainbow and brown trout species tend to take not only bottom-dwelling macroinvertebrates but also insects and crustaceans that swim or drift in the water away from the bottom. Trout and platypuses have co-existed in many bodies of water in Australia for over 100 years and it seems likely that dietary competition is minimised by differential use of the food sources in the streams occupied by the species. The consumption of trout eggs by the platypus has been reported in Tasmania and was also found during the winter in a feeding study in the Thredbo River in New South Wales. During winter, when platypuses must expend extra energy to regulate their body temperature, this source of food may be quite important where it occurs, although it presumably makes up only a small part of the diet in such streams. Obviously *Ornithorhynchus anatinus* is not totally dependent on trout eggs for its food in winter, as platypuses occur in areas where trout do not spawn, and they have been in Australia for much longer than the introduced trout, in fact, about five million years longer.

A number of native freshwater fishes, water birds and the native water rat (*Hydromys chrysogaster*) also consume macroinvertebrates as part of their diet. However, these species have coexisted in Australia's streams for thousands, perhaps millions of years, by partitioning their food resources. In its diet for example, the water rat takes items from the stream which do not appear to be taken by the platypus, including fish species, frogs and large freshwater mussels. The species is less dependent on water than the platypus, foraging along the stream banks for small mammals, lizards, water birds and fresh carrion. The water rat regulates its body temperature more poorly than the platypus and enters the water for shorter periods, particularly in the winter.

While a number of popular books suggest that the platypus eats prodigious amounts of food each day, reputedly consuming 50–100% of its own body weight in 24 hours, these claims warrant careful consideration. By weighing food supplied to, and retrieved from, platypus tanks in captivity and using calculations from the metabolism of the platypus in captivity and in the wild, several studies have shown that individual animals do not need to eat such large amounts of food. Actual food requirements were found to be around 13–28% of body weight each day, with animals having to consume more in the cold of winter and less in summer. However, during the period when a captive female at Healesville Sanctuary in Victoria had dependent young in a nesting burrow, she apparently consumed close to 100% of her own body weight per day during the later part of lactation (Chapter 2).

Body temperature regulation

In the late 1890s, a scientist reported a body temperature for the platypus of around 25°C. This observation led to the notion that the species was primitive with respect to its body temperature regulation and represented a stage somewhere between the 'cold-blooded' reptiles and the thermal competence of the mammals. But by the early 20th century, both naturalists and biologists had concluded that, while the platypus did have a low body temperature (32°C) when compared with other mammals, it was capable of maintaining this temperature over a range of air temperatures. In spite of this, the ability of the species to maintain its body temperature (homeothermy) in water was doubted and a prominent biologist in 1902 commented that '*Ornithorhynchus* is not an amphibious animal, and only goes into water for food or to amuse itself, and if kept too long in water its temperature falls and it dies'. The renowned Australian naturalist Harry Burrell, who spent inordinate amounts of his time observing platypuses in water both in winter and summer on the northern tablelands of New South Wales, agreed with this suggestion. However, more recent studies of captive and free-ranging animals showed that the species maintains a constant body temperature while actively foraging in water, even for extended periods in cold alpine streams, where water temperatures in winter may reach freezing point. In alpine areas in Tasmania, platypuses have even been observed swimming under ice and sliding over the snow.

During the very cold winter months of 1981, the waters of the Thredbo River reached freezing point and ice formed along its edges (see Colour Plate (top), page 36). The ponds in the Gaden Hatchery froze solid and only the flow in the stream itself prevented it from freezing completely. On the evening before the river froze for the first time that winter, a platypus was captured and a radio transmitter that measured its body temperature was fitted to its tail. It was kept overnight and, with some trepidation, biologists released the young male back into the partially frozen stream the following morning. This would have been the first time it had experienced ice, having only emerged from the nesting burrow in the February of that year. The coloured marking tape around the tail could be seen under the ice, as the animal swam out to the middle of the river, headed off upstream and disappeared. It quickly located a burrow in the river bank. As expected, the clicking of the transmitter indicated its body temperature remaining close to the normal level of 32°C while it rested in the burrow. A short time before dark the young male re-entered the stream and began to forage. The biologists, heavily clothed and huddling from the biting wind, listened in anticipation for a slowing of the click-rate of the transmitter signal, indicating a falling body temperature. No such change occurred, however, and during the eight hours the animal continued to forage in the near-freezing water, its body temperature fluctuated less than one degree either side of the normal 32°C – hardly the lifestyle of a 'cold-blooded' animal. This is patently a species able to regulate its body temperature even under extremes of ambient temperature.

When the amount of heat produced by an animal is the same as it loses to the environment, its body temperature remains constant. In the platypus, 25°C is the temperature at which its resting metabolism produces about the same amount of heat as it is losing to its surroundings. Body temperature can be maintained in a colder environment by elevating metabolism, producing more heat to compensate for the greater heat loss to the environment. When the environmental temperature of platypuses kept in the laboratory was reduced from 25°C to 5°C, metabolic heat production increased by 50%. When any animal is immersed in water, body heat is lost much more rapidly than it is in air. When immersed in water at 5°C, the heat production of these captive platypuses increased by more than three times what it was in air at 25°C. Compensating for heat loss by increasing heat production is costly in terms of energy expenditure and requires greater food intake. However, the platypus also has adaptations which reduce the amount of heat it actually loses to the environment, especially under cold winter conditions.

The fur, consisting of dense underfur and flattened guard hairs, provides insulation over most of the body that is comparable to 3 mm of neoprene wetsuit material. The fur is waterproof, having a hair fibre density of 600–900 hairs per square millimetre, which traps a layer of air that provides most of the insulation. The tail fur has few underfur fibres and has much poorer insulation than that of the rest of the body. Side views through the body and tail fur are shown in Figure 5.6. Often when a platypus dives (see Colour Plate (bottom), page 33), some of the air is squeezed out of the fur by water pressure but enough remains to keep the animal warm. Early naturalists, on seeing bubbles of air coming from diving platypuses, suggested the species was able to breathe through its back. As a result of their excellent insulating properties, platypus skins formed the basis of a hunting industry during the 19th century, when thousands were used to make hats, gloves and even rugs and bedspreads (see Colour Plate (bottom), page 36 and Chapter 8).

The platypus can also reduce heat loss by increasing its tissue insulation. Constriction of peripheral blood vessels, especially to the unfurred or poorly furred extremities – the bill, forefeet, hind feet and tail – is responsible for increased tissue insulation. The species also has a complex arrangement of veins and arteries in the pelvic region that forms a counter-current heat conservation system, called a *rete mirabile* (miraculous network). Warm blood, passing through the network of small arteries, exchanges a certain amount of its heat with the cooler blood returning to the heart via the parallel network of small veins. Similar counter-current systems are found in many species where unfurred or non-feathered extremities are exposed to cold environmental conditions. In Figure 5.7 the nature of this network is illustrated.

Figure 5.6 Platypus fur is extremely dense and provides excellent insulation. The depth from the skin to the tips of the guard hairs is around 1.5 cm. Left: dorsal body fur; right: tail fur. Photos: Bob McBlain

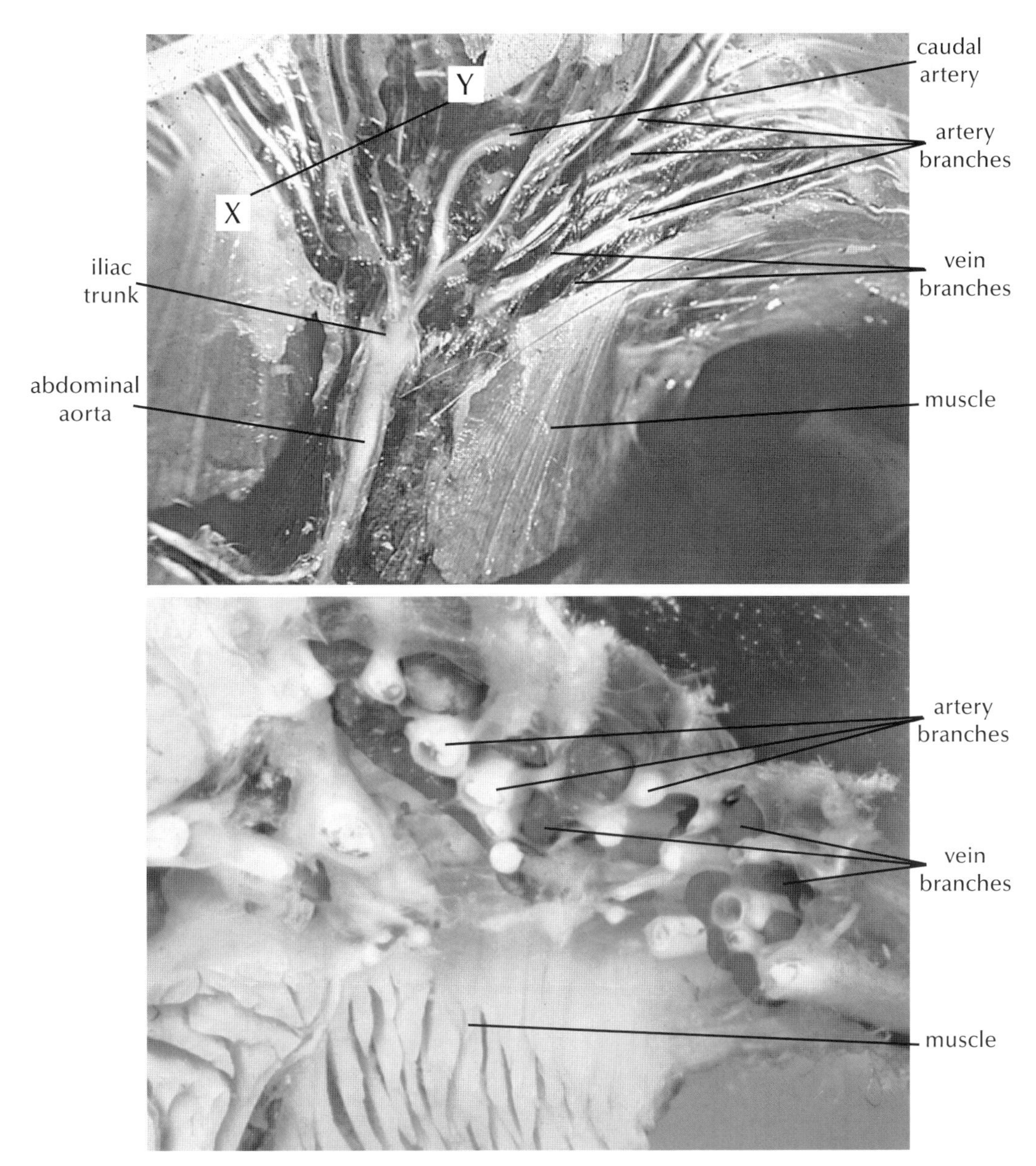

Figure 5.7 Counter-current heat exchange *rete mirabile* in the pelvic (thigh) region of the platypus. Top: The abdominal aorta (main artery) divides into two short iliac trunks before dividing into numerous branches supplying blood to the hind legs and thigh area (arteries are light coloured). Paralleling these arterial branches are the many veins returning blood to the posterior vena cava (main vein). Veins are dark coloured. The single caudal (tail) artery is also shown. Bottom: The section through the muscles of the thigh region shows arterial (light) and venous (dark) blood vessels positioned parallel to each other. Section is from X–Y above.

Many mammals that experience cold winter temperatures in their natural environment, including the short-beaked echidna, have a layer of fat under the skin as insulation. The platypus does not have such a subcutaneous fat layer but it does store fat in its tail. A platypus with good tail fat stores has a plump rigid tail, while one in poor condition has a strap-like tail. Sometimes the vertebrae are visible in an emaciated animal. In certain studies, it was found that fat stored in the tail increased over the warmer months of the year in some individuals, but was then mobilised as an energy source during the colder months, when energy demands were elevated. This trend is inconsistent, however, between study areas and in different years. Adult females and first-year males captured in the upper Shoalhaven River in New South Wales exhibited a change in body condition over the winter months but this trend was much less marked in adult males. Females rearing young often have low tail fat reserves at the end of the lactating period. The tail fat makes up around 40% of the total fat found in the body.

Exposure of animals to extreme air and water temperatures is also reduced by their use of burrows. In two separate studies, one on the mainland and the other in Tasmania, temperatures recorded inside natural and artificially constructed platypus burrows remained between 14 and 18°C while the outside air temperatures went from as low as –5.5°C to as high as 34°C. Unlike nesting burrows, which may be quite long and complex (Chapter 2), resting burrows are normally around one or two meters in length. These are constructed mainly in earth banks consolidated by the roots of riparian vegetation (Figure 5.8).

Platypuses have been known for some time to be stressed in temperatures higher than 25°C. Individuals in the wild normally avoid such temperatures, even in the tropical parts of their distribution, by occupying burrows where the insulation of the soil buffers the changes in ambient temperature inside the burrow. The loss of metabolic heat produced during exercise in water at temperatures higher than the normal body temperature of the species means that animals cannot tolerate water temperatures much higher than 25°C. Evaporative cooling is not available as a means of losing metabolic heat produced during exercise and heat loss by conduction is reduced by the small temperature difference between the platypus (body temperature 32°C) and the surrounding water. While suitable habitat seems to occur in the tropical north of Australia, especially Cape York, and along the streams traversing the inland plains of eastern

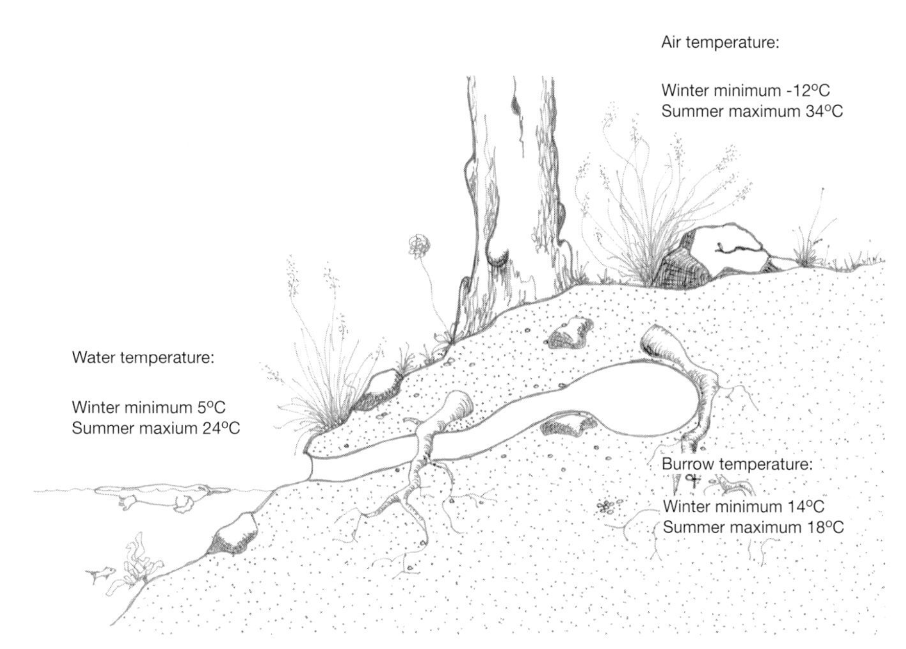

Figure 5.8 Burrows buffer against environmental temperature extremes.

Australia, platypuses appear to be absent from these areas. Thermal stress during foraging in warmer waters during summer may be a factor involved in this limit to distribution.

In the Snowy Mountains of New South Wales, the short-beaked echidna has been found to enter hibernation for several months during the winter. Echidnas, monitored with radio transmitters, lowered their body temperatures to between 4 and 9°C (close to the environmental temperature) from their normal active temperature of 28–33°C, and experienced spontaneous arousals, to around their normal body temperatures, for several hours every two to three weeks during the period of hibernation. The early naturalist George Bennett commented that platypuses could be seen in Australian rivers at all seasons of the year, but he observed that the species appeared more abundant in summer than in winter, and speculated on the possibility that individuals may 'not in some degree hybernate'. Harry Burrell also referred to periods of absence from the river, such as might occur during 'lethargy at times confused with true hibernation'. Observations have also been made of captive animals and radio-tracked animals in Victoria that remain in their burrows for periods of up to six and a half days, suggesting that they may have been hibernating or in a torpid state. Unfortunately the body temperatures of these captive and wild Victorian animals were not measured. In contrast to these observations, platypuses monitored during winter in the Shoalhaven and Thredbo Rivers in New South Wales and in Lake Lea in Tasmania, showed no evidence of hibernation or torpor and the possibility of torpor or hibernation during periods of inactivity still needs to be investigated.

Walking

Platypuses seldom travel overland, normally moving directly between the water and their nesting or resting burrows. Burrow entrances are usually at or within a few metres of the water's edge, although at Lake Lea in Tasmania, individuals were found occupying burrows at a distance of up to 20 metres from the water. At times, platypuses are also reported moving considerable distances on land, especially in drought and during dispersal of juveniles. At these times individuals (often first-year juvenile males) are found in drains, farm dams, tanks and other unlikely places, including sewage treatment ponds and mine tailings dams. So is the species good at walking? The answer to this question unfortunately seems to be a resounding 'no'! The limbs of the platypus, which are heavily built and more adapted to digging than to walking, are somewhat splayed away from the body. The body is not continuously held above the ground surface and the energy required for walking is 19–27% higher than for most terrestrial mammals of similar size. The average platypus can outsmart a pursuing biologist careless enough to let one escape, but in this instance the maintenance of liberty is due to swerving and weaving and not to sheer speed over land. Pursuit by a skilled predator, such as a fox, would in most instances have a different outcome (Chapter 8). Interestingly, until recently there were no foxes in Tasmania, where longer distances between water and burrows have been observed. Other aspects of platypus ecology, including activity, movements and home ranges, are discussed further in the following chapter.

6
ECOLOGY

'Soon the river was before me, the banks of which were adorned by pendulous Acacias, which at this season of the year [September] were profusely covered with their rich golden and fragrant blossoms, while the lofty majestic Eucalypti or Gum-trees, many of which were young and gracefully pendent, together with the Swamp Oaks or Casuarinaea, resembling firs at a distance, added to the variety and natural beauty of the landscape'. ... 'The sun was now near its setting, when, at a more quiet part of the river (knowing, as I did, the crepuscular nature of the animals) I endeavoured to obtain a sight of the shy Ornithorhynchus paradoxus*'. ... 'At a tranquil part of the river, called by the colonists a "pond", on the surface of which numerous aquatic plants were growing profusely, or in places of this description, the Water-Moles were most commonly seen, seeking their food among the plants, whilst the shaded banks afforded them the excellent situations for excavating their burrows'.*

GEORGE BENNETT (1860)
Describing a typical river on the southern tablelands of NSW in the 1830s

If George Bennett were to describe the same stream on the southern tablelands of New South Wales today, it would almost certainly be different. The wattles, with their 'rich golden and fragrant blossoms' and the 'majestic Eucalypti', would probably not feature in the description or they would be there in reduced numbers. The 'Swamp Oaks' (*Casuarina cunninghamiana*), however, may still line the now narrow strip of riparian vegetation along the stream bank and the 'shy *Ornithorhynchus paradoxus*' [*anatinus*] would still be likely to occupy the stream.

Distribution and abundance

The overall distribution of the platypus in Australia shown in Figure 1.6 is a broad brush-stroke representation of the real world. Platypuses obviously are not found throughout the total area shaded on such a map. The platypus is dependent on water. The water bodies of Australia form an irregular mosaic across the landscape, and so the distribution of the platypus is also discontinuous within and between river catchments in its overall distribution. At the catchment scale, the general feeling of confidence regarding the current distribution of the platypus gives way to thoughts of knowledge gaps and areas of concern.

On a trip to the western settlement of Bathurst in 1815, Governor Lachlan Macquarie reported 'great numbers of water moles in the Campbell-River at Mitchell Plains', and two were brought back to camp from the Fish River after an evening's fishing by one of his party. The Scottish scientist, William Caldwell – the first European to confirm that the species laid eggs – also indicated that platypuses were 'very numerous' during his work in 1884 in the Burnett district of Queensland. The German biologist, Richard Semon, wrote in 1899 that he caught 'a considerable quantity' in the same area in winter, although he had less success during summer, when he suggested they possibly emerged when it was too dark for hunting. Although he described the habitat, behaviour and other aspects of the biology of the platypus, George Bennett made no comment on the abundance of the species in the Goulburn, Yass and Tumut areas of New South Wales in the 1830s. Whatever their abundance before European occupation of Australia, platypuses are reputed to have declined in numbers during the days when their furs were traded, prior to their protection in all states by 1912. Since then, population numbers were reported to have increased by around the 1940s but there is no objective evidence available to indicate that the suggested increased abundance at that time translates into current abundance being similar to that in pre-European Australia. In his book, published in 1927, Harry Burrell described a report of the apparent migration of 'at least' a hundred platypuses along the

Gwydir River after a flood in 1859, but no such phenomenon has ever been recorded since.

Using mark and recapture data, a minimum population of 14–18 individuals was estimated to be occupying two large river pools connected by a long riffle sequence covering approximately 1.5 km of the upper Shoalhaven River (9.3–12.0 platypuses per km). This estimate is very high compared to those made in smaller streams near Melbourne and on Kangaroo Island and may be an overestimation of the population as a result of a number of factors related to the assumptions that are made when calculating population size using mark and recapture methods. Some examples from field work illustrate a few of the limitations of mark and recapture studies.

One method of catching platypuses to be marked and released is to use unweighted mesh ('gill') nets that entangle animals but permit them to come to the surface to breathe (Figure 6.1). This method is only suitable in stationary pools or slow-flowing water. Where there is a current, the nets rise off the bottom, permitting foraging platypuses to pass under them. Nets are placed parallel to the flow of the stream, to reduce material becoming entangled. This configuration has also been found to result in better capture rates than nets set across the stream, but by setting nets this way some animals avoid being captured. There also appear to be differences in the ease of capture between individuals in the population – some females are captured many more times than others and males are recaptured much less often than females. For example, during a study of body temperature of radio-tracked platypuses in the Thredbo River in the Snowy Mountains area of New South Wales, the pools in which the animals spent most of their time during the study were well known. However, after the failure of certain radio-tags, individuals marked with these tags were not able to be recaptured in these same pools, despite considerable netting effort. In addition, netting in artificial ponds within the confines of Warrawong Sanctuary in South Australia also failed to capture a male platypus known to be in the sanctuary, as it had sired young in the captive population. Mark and recapture studies have also shown considerable mobility by certain individuals in platypus populations. Both unequal ease of capture and movement in and out of an area can affect the accuracy of population estimates using mark and recapture methodology. Normally, to make a population estimate, it is assumed that another sample of animals caught after those initially captured and marked will represent the proportion of marked to unmarked animals in the whole population, i.e. if a population is very large, only a small proportion of it will have been initially marked. If animals previously captured and marked go out of their way to avoid the nets (often referred to as 'trap shy'), the next sample will be biased towards unmarked animals and an overestimate of the whole population may occur. Conversely, if most of the animals initially marked spend the majority of their time in the sampling pool, and there are others which use the pool, but only infrequently, it is more likely that marked ones will be recaptured. As a result, there will be a high proportion of marked individuals compared to unmarked in the next sample and an underestimate of the population could be made. However, although an accurate estimate of the population may not have been possible, the platypus population in the upper Shoalhaven River is almost certainly much higher than that in other smaller streams where population estimates have been made. Since 1973, when the study began, 327 platypuses have been captured in the most-studied 1.5 km section of the river and 714 individuals have been caught in 16.5 km of stream.

In several small streams near Melbourne, biologists from the Healesville Sanctuary and the Australian Platypus Conservancy have

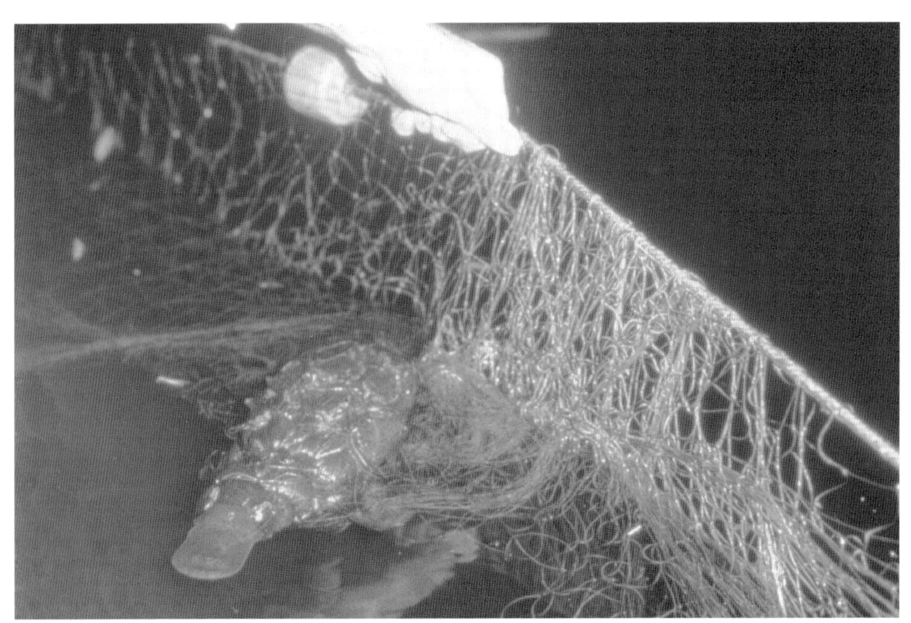

Figure 6.1 Platypus caught in an unweighted mesh (gill) net. Unweighted nets are used in field studies to permit animals to reach the surface to breathe. Illegally used nets often drown platypuses. Photo: Richard Whittington

used a series of fyke nets (Figure 6.2) to capture foraging platypuses during radio-tracking studies. These nets consist of a wind-sock shaped bag with several constrictions (or 'valves') preventing the escape of platypuses which enter the nets. The tops of the hoops supporting the netting, and the end of the net itself, are kept above the water level to provide an air space permitting trapped animals to breathe. Using this method, which is most suitable for small streams, mesh wings from each net are stretched across the stream to direct foraging platypuses into the nets, which are set in pairs with one facing upstream and one downstream. Assuming that the method captured the entire population moving through or within the study area at the time, population estimates of 1.3–2.1 adult individuals per km of stream were made. In the Breakneck and Rocky Rivers on Kangaroo Island, where the platypus population was introduced, the same methodology was used and population estimates between sections of the rivers and between seasons ranged from 1.3 to 5.2 platypuses per km. Although it provides a useful estimate of density per km of stream, this method may not always yield accurate population estimates since platypuses are known to leave the water to move around obstructions.

Figure 6.2 Fyke nets are also used to trap platypuses. The end of the net and the hoops supporting the netting material must be kept out of the water to permit animals to breathe, otherwise they will drown. The 10-cm square by-catch reduction grid across the opening was unsuccessful in preventing platypuses from entering this net.

An often-posed question is 'how many platypuses are there' in this stream, area, state or the whole of Australia and the response must always be a tentative one. The late Merv Griffiths, who worked extensively on various aspects of the biology of both species of Australian monotremes, mischievously said he frequently gave any number 'off the top of his head' in answer to that question but noted it was always best to give an odd number. He suggested an odd number was more likely than an even one to be accepted as an authoritative response!

Several biologists have used visual observations to assess the presence or absence of platypuses in a given area. The numbers of individuals observed at one time can be used as a measure of broad categories of abundance, such as 'rare', 'common' or 'abundant' or even as a numerical value, such as numbers of individuals seen per unit of observation (e.g. numbers seen per hour or per km). However, observations and netting done simultaneously show that not seeing platypuses may not necessarily indicate absence of the species from an area. Platypuses have been captured at a number of locations where local people have claimed never to have seen them. Conversely, observations have identified the presence of platypuses in places where netting only on one or two occasions has failed to capture the species.

Accurate quantitative data on population sizes are no more available now than they were in the times of the early naturalists, but netting, observations and community based questionnaire surveys suggest a reduction in the current local distributions and numbers of platypuses in particular streams (Chapter 8). However, research also indicates that the platypus may have been more abundant prior to European colonisation but occurred over much the same overall distribution. The exception to this is the Mount Lofty Ranges and Adelaide Hills of South Australia, where the species seems to have disappeared.

Habitat requirements

When not foraging, the platypus normally spends most of the daylight hours tucked up inside a burrow in the bank of the creek, river or pond (Figure 5.8). However, some individuals also occasionally use rocky crevices or piles of stream debris (see Colour Plate (top), page 40) as shelters or burrow under the roots of vegetation, sometimes several metres from the stream.

Observations by naturalists and more recent field research studies have identified a range of habitat variables associated with the presence of established populations of platypuses. A 'platypus heaven' is a river or stream, much like that described by George Bennett in the 1830s, with earth banks consolidated by the roots of native vegetation. The foliage of this vegetation at shrub and tree height provides shading of the stream and cover near the bank, where platypuses may forage and where they enter and leave both nesting and resting burrows. Many stream habitat features influence the production of macroinvertebrate species, which are the major food resource of the platypus (Chapter 5). These include the presence of logs, twigs, roots and instream vegetation, as well as the type of substrate on the bottom. Higher macroinvertebrate productivity is usually associated with areas where logs, roots and vegetation provide a range of habitats and food for these species and where a cobbled or gravel substrate provides fixed habitat, rather than shifting substrates, such as sand. The complexity of benthic habitats and the presence of aquatic vegetation have also been shown to be indicators of the occurrence of platypuses, which feed both in pools and riffles. The presence of pool–riffle sequences is also often associated with the occurrence of the species. Although these attributes are considered to result in 'ideal' habitat, the species is found in places where not all of these habitat variables are found, including streams considered moderately degraded by various human activities. The platypus appears to have coped with considerable habitat change over much of its range but thresholds for survival and reproduction in such degraded habitats are not known (Chapter 8).

Mobility, home range and juvenile dispersal

It was just after dark as male #56 slid into the water of his kilometre-long home pool on a cool autumn night. In spite of the small hindrance of having a transmitter package attached near the base of his tail he made his way upstream from one of his resting burrows, diving effortlessly for food in the cold black water. First he moved along the bank, diving around the snags and overhanging willows before gliding under the water to a three-metre deep area of the pool, where he searched for freshwater shrimps. These normally moved about in the water column itself, but in cold conditions are found on the bottom. As he dived and surfaced he probably was aware of several trout also feeding in the area, and sensed a large eel nearby. He reached the rapids at the top of the pool and moved through the rocks of the shallow riffles to get to the next pool upstream. On a moonless night like this he was probably fairly safe from any foxes which may have been wandering near the river. This accomplished and finding little to his liking in the next pool he moved through two more riffles to forage successfully in a pool much further up the river. It was now just after midnight and he had covered about two kilometres of river. It was time to go back and he swam quite quickly, only occasionally diving to pick up a tasty morsel of food here and there. He swam down through the other pools, rapidly scrabbling around the rocks of the riffles and back to the large pool. He could have gone into a resting burrow he had used before at the top of the pool, but still being a little hungry spent the next few hours zigzagging in a leisurely manner between good feeding areas in this pool, which was the nucleus of his home range. Around six-thirty in the morning a slight pink glow above the hills to the east was beginning to appear and he returned to his burrow to sleep.

The narrative above describes the anthropomorphic musings of a biologist following a radio-tagged platypus during a feeding excursion along the banks of the upper Shoalhaven River in New South Wales on a cold and moonless May night in 1980. Male #56 had been in the water for just over 13 hours, when it finally retired to one of its burrows to rest. Platypuses normally spend between 10 and 12 hours foraging each day (Chapter 5). This can vary from season to season and is sometimes split into separate periods, rather than just one long one. While the maximum distance known to have been covered by a platypus in a 24-hour period is 10.4 km, many will forage within a single pool, presumably depending on the availability of macroinvertebrate food sources, which are often patchily distributed.

Over longer periods of time, adult platypuses have been found moving over distances of up to 15 km, but in a study in the upper Shoalhaven River in New South Wales only 32% of platypuses recaptured were caught outside the pool in which they were originally netted, with less than half of these moving over distances of more than 1 km. However, other mark and recapture and radio-tracking studies have shown that at least some individuals in most populations are more mobile and include a number of adjacent pools in their home range; the part of their habitat in which they spend most of their time feeding, breeding and caring for their young. In the Thredbo River in the Snowy Mountains, radio-tracked platypuses exhibited variable home ranges, often using a number of resting burrows in the same or different pools in the area. The variation in home range sizes and locations of burrows within these ranges during the study are shown in Figure 6.3. Platypuses in several streams in Victoria also had similar home ranges to those in the New South Wales rivers (Table 6.1). In Lake Lea, a small sub-alpine lake in the Cradle Mountain area of Tasmania, the area of the lake used by radio-tracked platypuses varied greatly between individuals, some covering as little as two hectares of the lake and others up to 58 hectares.

Table 6.1. Recorded home ranges and maximum distances covered by platypuses in a number of radio-tracking and mark and recapture studies.

Home range (km)	Location
0.2–2.0	Upper Shoalhaven River, New South Wales
0.4–2.3	Thredbo River, New South Wales
0.2–2.3	Badger Creek, Victoria
2.9–7.0	Badger Creek and Watts River, Victoria
0.4–2.6	Goulburn River, Victoria
2.9–7.3	Yarra River, Mullum Mullum and Diamond Creeks, Victoria

Several studies also suggest that, although home ranges may overlap, especially those of female, sub-adult and juvenile animals, there may be less overlap of home ranges by males.

Mark and recapture studies in both Victoria and New South Wales have found that, after they emerge from the nesting burrows in January to March, juvenile platypuses can be captured for a number of months in their 'home' area, but then the numbers of recaptures of the majority of these animals decline. This may occur as early as the middle of the year they first become independent. Some stay longer, but most have left their 'home' area by the end of their first year of life. Juveniles may disperse in search of new places to live. It is even possible they may be forced to do so by competition for food or burrows with the resident members of the population, which may include their parents. Many juvenile platypuses caught and marked in an area are never captured again. Of 137 juvenile females and 94 juvenile males caught in the upper Shoalhaven River in New South Wales, only 32% and 14% respectively were recaptured one or more times. Based on reports of such low recapture rates of juvenile males, and recorded cases of one male spurring another to death during close confinement in captivity, it has been suggested that in the wild juvenile males

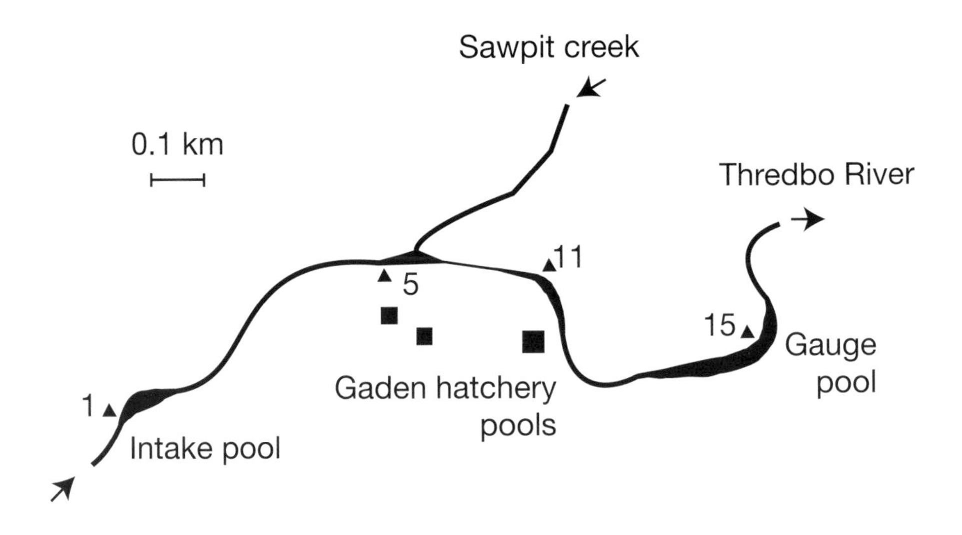

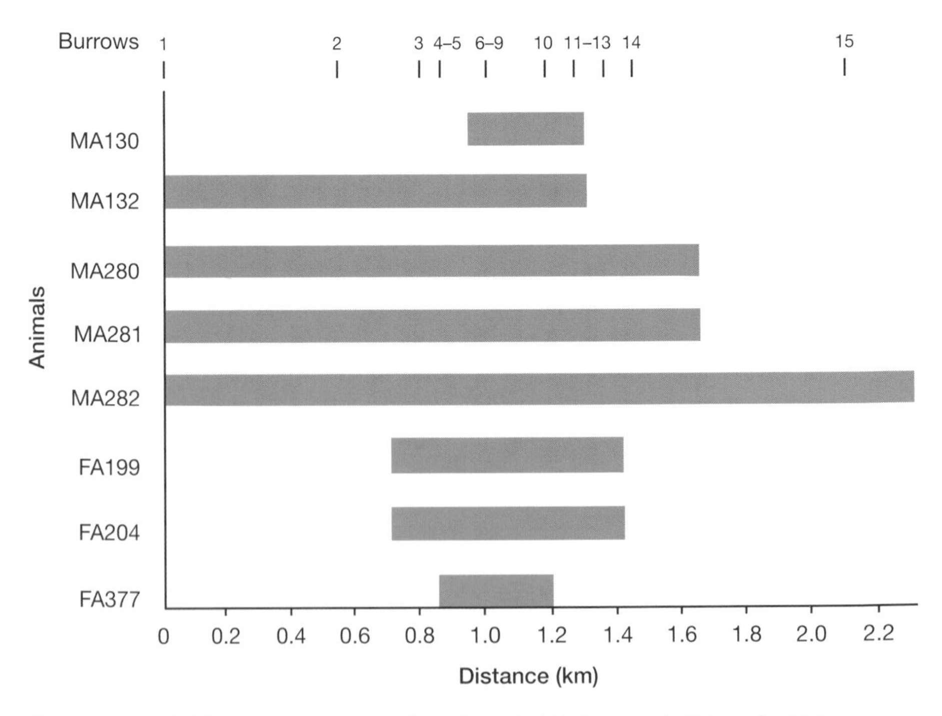

Figure 6.3 Individual home ranges of adult male (MA) and adult female (FA) platypuses radio-tracked in the Thredbo River in New South Wales were of variable length (horizontal bars). These home ranges overlapped each other and several of the 15 burrows (numbered triangles) located during the study were used on different occasions by several individuals.

may be killed by adult males. However, to date no data have been published to substantiate this notion.

Male juvenile and sub-adult platypuses have been found to cover longer distances than those travelled by adult animals. In the Shoalhaven River a newly independent juvenile male was followed downstream more than 3.5 km over a period of two days before it lost the transmitter. During studies by the Australian Platypus Conservancy in Victoria, one microchipped juvenile travelled at least 40 km in an 18-month period in the Yarra River system, another from the Melbourne area moved 43 km over two years and nine months and a young male in the Wimmera River system covered 48 km in just seven months. It is not known where dispersing juvenile platypuses finally end up, and further long-term radiotracking is really the only way to find out.

However, some juvenile platypuses do remain in their 'home' area but apparently do not have an easy time. Juveniles recaptured after their first winter in the upper Shoalhaven River were found to have lost a lot of weight and to have considerably depleted fat reserves. During the Tasmanian winter of 1990 three dead juvenile platypuses were handed into a veterinary laboratory near Launceston. All had no fat reserves in their tails and one had pneumonia. Adult platypuses may lose weight during winter, but the losses in juveniles appear to be more dramatic. Platypuses are often reported in odd places such as farm dams or drains and even being killed on roads that are quite a long way from water. A few of these animals have been positively identified as juveniles by reference to the stage of development of their spurs (Chapter 3). Perhaps many of the individuals found in places that are less than adequate for their survival are dispersing juveniles moving out from the more preferred habitats already occupied by resident animals.

Drought and flood

Australia is well known as a land of drought and flood, although in recent years droughts have been more prevalent than floods, as a result of repeated El Niño events and global climate change. However, even during otherwise dry times, floods can occur as a result of heavy local rainfall events in the south-eastern states, monsoon events in the north and even melting snow affecting the waters of the west- and north-flowing streams around the Snowy Mountains and Victorian Alps. For example, in late

October 2004 steady rain for several days in its headwaters resulted in the Hastings River on the north coast of New South Wales rising from a low flow of only a few hundred megalitres per day (one megalitre is equivalent to about half the volume of an Olympic-size swimming pool) to several thousand. Figure 6.4 shows a pool-riffle sequence in the river during and after the flood. Platypuses were observed in the pool as the floodwaters began to rise and again after the stream had returned to its pre-flood levels.

Although few incidents come to notice, floods may cause significant mortality in platypus populations. During the late spring of 1981, a female platypus was handed in to a local veterinarian in the Wollongong area of New South Wales. The unfortunate animal, which later died of pneumonia, had been washed out of its burrow when a flash flood in Yellow Rock Creek destroyed many of the creek's banks. This female must have had young in the nesting burrow, as she was producing milk at the time she was picked up in an exhausted condition by a farmer's wife. Judging from the extent of the destruction of the banks along this creek, it is probable that this animal was not the only one to be washed out of its burrow. Dead platypuses have also been reported in several other New South Wales rivers after floods, presumably drowned in the rapid flows, struck by floating objects or entangled in debris carried by the stream (Figure 6.5). Mortality of burrow-dependent young may also occur during flash flooding. Although the females apparently block the burrows with earth plugs, these may be insufficient to prevent the entry of water and young may drown. In the 1991/92 breeding season in the upper Shoalhaven River in New South Wales, 50% of the females captured in December were lactating and therefore supporting nestlings. During sampling in early March, however, when the nestlings should have been leaving the burrows to become independent, no juveniles were captured. It is suspected that two sudden short-duration flood events in late December and early January may have resulted in high nestling mortality. Early in February 2005, the Melbourne area received more than 120 mm of rain in less than 24 hours. After the flood which followed this rainfall, biologists from the Australian Platypus Conservancy recorded a capture rate of only 0.03 juveniles per net site per night; one-tenth of the average juvenile capture rate recorded in these streams from 2000 to 2004. This observation suggested that a number of the newly independent juveniles had been drowned, entangled or struck by debris.

Figure 6.4 A flood in the Hastings River in New South Wales forced the abandonment of a platypus study at the time (top) but platypuses were again observed in the pools when normal flows returned (bottom).

Figure 6.5 Stream debris along the banks of Jerrabattgulla Creek on the southern tableland of New South Wales show the height of a flood thought to have drowned a number of nestling platypuses during the 1991/92 breeding season.

In the first five years of a 30-year study in the upper Shoalhaven River in New South Wales, seven major floods occurred which changed the river from its normal series of deep pools with connecting riffles into a raging mass of brown water with no clear distinction between pool and riffles. Marked animals, however, continued to be recaptured after these events. For example, of eight platypuses caught following one of these floods, the peak of which was over eight times the normal level, five had been marked in the area previously and two of these had been captured in the same pool immediately before the flood. Although one dead animal was found on the bank after a flood, suggesting some mortality, it appears that many individuals are not even displaced from their home ranges by rises in river level.

The platypus has occupied the rivers of the Australian mainland and Tasmania for around five million years and has presumably evolved strategies to cope with prevailing river conditions, including flooding. However, how platypuses cope with natural floods is unknown. Early naturalists suggested that they occupied rabbit burrows and hollow logs away from the river and fed in backwaters and billabongs during floods, returning when the waters had subsided. Recent anecdotal observations

Figure 6.6 Like this one, most riffles in the upper Shoalhaven River in New South Wales were dry during the severe 1979–83 drought. Because of its dependence on water in which to feed, prolonged drought conditions are of great concern for the conservation of the platypus.

seem to support these ideas, as did a radio-tracking study of male platypuses in the Goulburn River downstream of Eildon Weir in Victoria. In this study, during high flows from the release of irrigation water (essentially an artificial flood), platypuses avoided the faster-flowing waters and moved to several shallow ponds that were normally separated from the river when water was not being released from the dam. Interestingly enough, the home ranges occupied by these males did not increase or decrease during river flow changes and the activity periods remained very similar, as did the numbers of occupied burrows.

In the northern part of its range, the platypus experiences floods during the summer monsoons ('wet seasons'). Little is known of the biology and distribution of the species in this part of Australia. Unfortunately, a radio-tracking study undertaken in the Barron River on the Atherton Tablelands in the late 1990s to investigate responses of platypuses to high wet season flows was plagued with problems, including transmitter malfunction and then by atypical wet season conditions with minimal rain. Platypuses were, however, marked and released in this study and were again found in the same sections of river before and after flood flows, including those associated with Cyclone Rona in 1999. Platypuses were seen foraging in the slower-flowing water along the banks of the flooded river but they also entered the faster-flowing water at times. It is still not known if platypuses in fact 'go with the flow' and then return, or if they continue to tenaciously occupy their normal home ranges during floods.

Droughts are, of course, another matter. The platypus is highly specialised and dependent on obtaining its food from the bottom of the water bodies in which it lives. For this reason, the most recent sequence of droughts, and the apparent probability of more prolonged drier conditions over at least some of its range, is of concern. Currently it is presumed that in good years, many of the juvenile platypuses recruited to populations move out to occupy the more marginal areas that have suitable habitat, but in bad years these animals probably suffer considerable mortality. During drought events platypuses are more often reported being killed by terrestrial predators, particularly foxes (Chapter 8), probably as a result of having to scramble through shallow or dry riffle areas when moving from one remaining pool to another.

Most riffle areas connecting pools in the upper Shoalhaven River in New South Wales were completely dry for weeks at a time during the severe drought of 1979 to 1983, when even the larger pools were also reduced in size. Catching relatively large numbers of platypuses was fairly easy during the biological studies done at that time, as the animals were concentrated in the large pools, which acted as feeding refuges. After that time, many pools in the area began to fill with sand as a result of damage by cattle and wombats and intermittent high flows. During the subsequent shorter drought event of 1994–1995 it was expected that animals would be in poorer condition and that breeding would perhaps be impaired as fewer deeper refuge pools would be available. That did not happen. Even during the most recent extended dry or drought conditions, beginning in 2001, platypuses were still being captured in the upper Shoalhaven River study area, successful breeding was occurring and the individuals captured were not in poor or emaciated condition, as had been anticipated. Biologists from the Australian Platypus Conservancy also found capture rates and body condition of platypuses little changed in the Mackenzie River in the Wimmera district of western Victoria during droughts. How does this small and highly habitat-specialised mammal cope with such adverse conditions? The answer to that question may be its mobility. Animals may forage over much greater distances, find and occupy areas that retain refuge pools, or include such refuge areas in an expanded home range. Platypus captures during non-drought seasons in the Mackenzie River occurred along the entire length of the river in the wetter winter and spring periods but were concentrated in sections of the stream that retained more water during the drier summer and autumn. Perhaps searching for such refuges is also important for survival in drought times, when suitable habitat shrinks and numbers become concentrated.

Such a strategy would enable the species to cope with cycles of dry and wet times but survival during extended droughts may be another matter. The fossil platypus species almost certainly occupied wetter areas of Australia which dried out as the continent drifted north to its current position, and it may well have been these changes which led to their extinction (Chapter 7).

7
ANCESTRY AND EVOLUTION

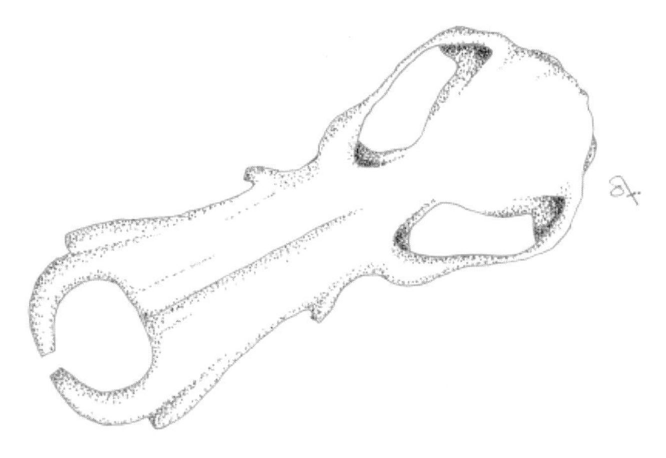

'The living platypus, sole survivor of a group that was once widespread, shows all the signs of having crept too far out on its thinning branch of the tree of life. In the past 15 million years alone, it has changed from a robust, toothed mammal spread over much if not all of Australia, into a small, toothless creature confined to a few east-coast river systems'.

MIKE ARCHER (1995)

The fossil record seems to suggest that the platypus is the sole survivor of a previously more diverse group of platypus-like monotremes that has occupied aquatic habitats in Australia, South America and probably Antarctica from at least as early as 120 million years ago (early Cretaceous).

The platypus family

Platypuses have teeth as nestlings but these are replaced by horny grinding pads made of keratin soon after the young leave the nesting burrow and begin to consume food from the bottom of the river. However, before these teeth are lost they are very distinctive, having a unique pattern of ridges and grooves on their chewing surfaces and six separate roots. In 1971 in the southern Simpson Desert area of South Australia, where today there is little water, let alone platypuses, palaeontologists (scientists who study fossils) found a fossilised part of a hip bone (ilium), lower jaw (dentary) fragment and a molar tooth with features characteristic of those of modern nestling or juvenile platypuses. Soon after, in desert country east of the Flinders Ranges in South Australia, another one of these teeth was discovered. All of these finds were eventually attributed to an ancient species of platypus, to which the name *Obdurodon insignis* was given. Its age was determined to be around 24 million years. It was thought to have had a shorter bill than the modern species. Other materials found with these *Obdurodon insignis* fossils, such as turtle bones and the teeth of lungfishes, suggested that it probably lived in or near water. This new species was placed into the family Ornithorhynchidae, which up to that time had only one member – the modern platypus, *Ornithorhynchus anatinus*.

More than 10 years later, a small fragment (3 cm long) of opalised dentary bone, with three similar teeth in place, was discovered by miners David and Alan Galman at their Lightning Ridge opal mine in New South Wales. This tiny and stunningly beautiful fossil was identified as another platypus-like monotreme, but was considered to be different enough from the earlier fossils to be placed in a separate family. It was bought by the Australian Museum in Sydney and named *Steropodon galmani*. The fossil was estimated to have come from the early Cretaceous period, around 110 million years ago, when dinosaurs dominated Australia and the rest of the world. The platypus-like features of the fossil suggested that the whole monotreme group must have originated further back in time than the age of the fossil itself. By the time *Steropodon galmani* lived in Australia, the large southern continent of Gondwana had already undergone fragmentation. Africa and India had separated and were moving north but Australia was still firmly attached to Antarctica, which in turn was connected to South America.

Most palaeontologists had long been convinced that the diverse marsupial fauna of Australia had almost certainly originated in South America and so also began to wonder where monotremes originated. As a result, there was great excitement when in 1991 a distinctively monotreme fossil tooth was found in Patagonia in Argentina. Its owner was named *Monotrematum sudamericanum*, and because the tooth was very like those of *Obdurodon insignis*, the new species was also placed in the family Ornithorhynchidae. *Ornithorhynchus anatinus* now had two other members in its family and a platypus-like relative in a related family of monotremes.

The South American tooth was estimated to be around 62 million years old. Tantalisingly, two more of these teeth were discovered in the same area a year later. However, no other material has been found since that time and the anatomical details of the South American member of the family remain a mystery.

A year after adding the South American platypus to the family, another fossil was described from the famous Riversleigh World Heritage site in north-western Queensland. This specimen, actually discovered in the 1980s, was an intact skull of a very large platypus, complete with specialised upper and lower jaw bones modified into a bill (Figure 7.1). It was found in a fossil bed along with many other freshwater aquatic forms, including several species of freshwater turtles and fishes, indicating that the animal probably had an aquatic or amphibious lifestyle. The teeth were larger than, but very similar to, those of *Obdurodon insignis* from South Australia. The new species was placed in the same genus but given a different species name – *Obdurodon dicksoni*. This new species was apparently alive around 15 million years ago and was obviously specialised for feeding in water. Unlike the modern platypus, the adult animal had retained its teeth, suggesting that it may not have fed on small macroinvertebrates but on larger prey items. Two molar teeth of another fossil species were later found at a site in central Australia and, although not fully described, these also have been attributed to the genus *Obdurodon* but not yet allocated to a different species. So now three platypus species had become four or possibly five.

In 1995, jaw fragments and teeth from two more fossils were found. Both were around 100 million years old – one from Lightning Ridge in New South Wales (*Kollikodon ritchiei*) and the other from a marine rock platform in Victoria (*Teinolophos trusleri*). Although different from any fossil previously found, they were thought to have been platypus-like in basic form and to have lived in aquatic habitats. These species were different enough from each other and from the species attributed to the platypus family for them to be placed into two separate families. *Kollikodon ritchiei* has been placed into its own family (Kollikodontidae) and *Teinolophos trusleri* into the same family as *Steropodon galmani* (Steropodontidae). The increasing diversity of these fossil monotreme discoveries continues to support the hypothesis that the modern platypus is the sole survivor of a much more diverse range of platypus-like ancestors once occupying Australia and the rest of the eastern part of the Gondwanan land mass.

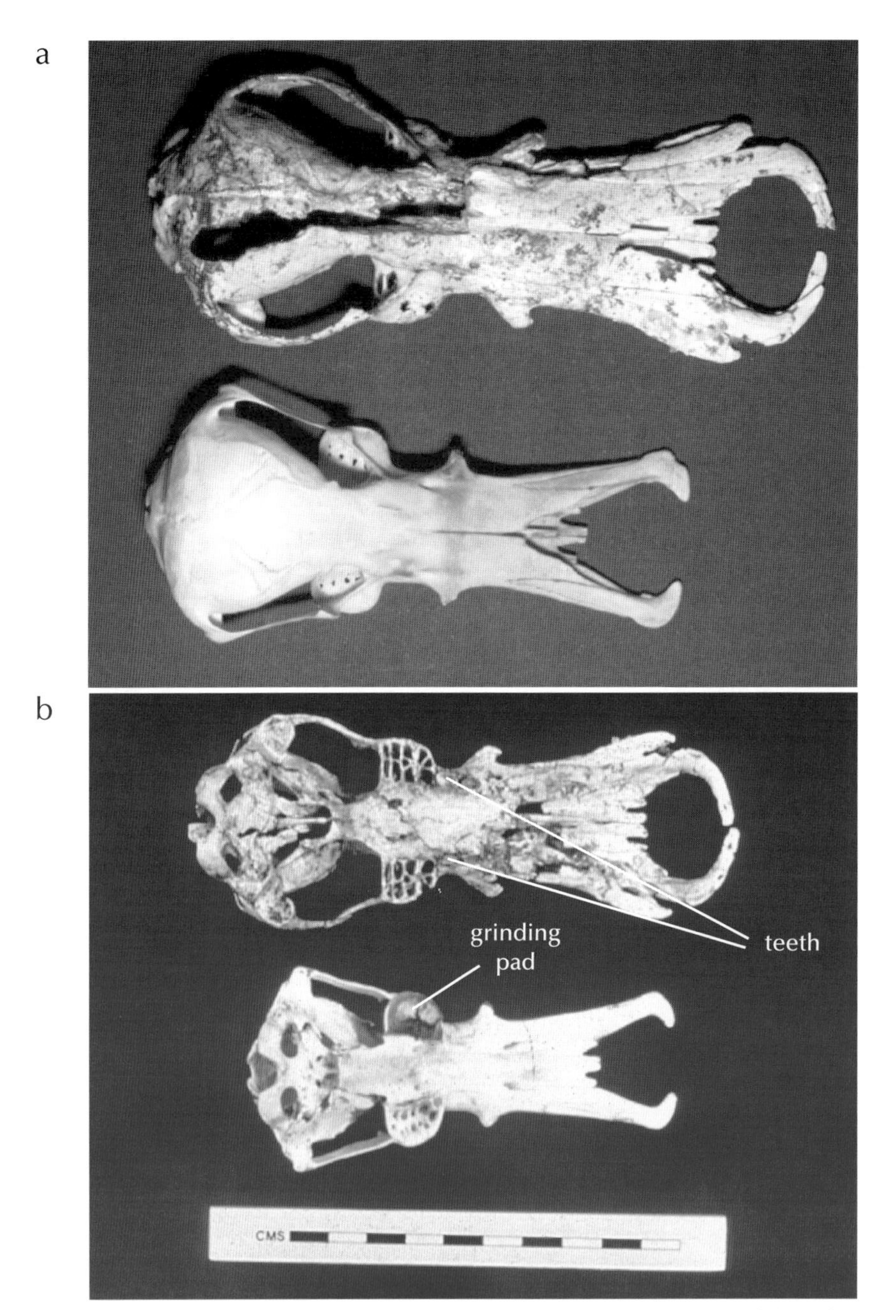

Figure 7.1 The large skull of the 15-million-year-old fossil platypus *Obdurodon dicksoni* (top) compared to that of a modern day platypus (bottom). (a) Dorsal view; (b) Ventral view showing adult teeth still in place in the fossil skull and the horny pad (missing from one side on this skull) that replaces the teeth in the modern platypus. Photos: Jenny Taylor/Riversleigh Project

A painstaking and arduous area of research, palaeontology is analogous to looking for a needle in a haystack. Even then, in many searches only part, rather than the whole needle, may often be found. On closer examination of that part, or with the finding of another part, it may be decided that in fact the whole object discovered was actually a safety pin rather than a needle. For example, the relationships of the fossil *Kollikodon ritchiei* from Lightning Ridge, which was thought to have been a platypus-like monotreme, are currently a matter of increasing debate. Its unique teeth, which look like the top of a hot cross Easter bun, have led a number of palaeontologists to question whether the species might not have been a monotreme, but rather a representative of a separate group of mammals altogether.

Table 7.1 shows the currently held view of the relationships between families and genera of monotremes, including as many as seven living or fossil species of echidnas, which are consigned to three separate genera.

Platypus family origins

Although not strictly considered part of the platypus family, fossils of the closely related platypus-like Steropodontidae (*Steropodon galmani* and *Teinolophos trusleri*) began to appear towards the latter part of the early Cretaceous period around 100 million years ago. Because the fossil material for both of these species indicates considerable specialisation, palaeontologists suggest that this group would have originated almost certainly prior to the Cretaceous period, possibly during the Jurassic (203–135 million years ago) or even in the late Triassic period (over 203 million years ago). Unfortunately the Jurassic period in Australia has yielded little or no mammalian fossil material and is considered to be a palaeontological 'dark age'.

The platypus family, or its near relatives, appears to have originated at least 100 million years ago. But did the group arise in the Australian or South American parts of eastern Gondwana? The diversity and more widespread distribution of fossil material from Australia, compared to the three teeth of one member of the immediate family found in South America (*Monotrematum sudamericanum*), suggest to palaeontologists that platypuses possibly came from the region that is now Australia, although further research in South America and possibly Antarctica may end up refuting that hypothesis.

Table 7.1 Relationships of the described families, genera and species (number of species from each period is shown) of the Order Monotremata. Based on living (normal font) and fossil (**bold font**) genera/species. The geological periods in which these specimens were found are also shown. No monotremes have yet been found from the late Cretaceous, Eocene or late Miocene. *Classification of the *Kollikodon* species as a monotreme is currently the subject of debate. Modified from *Prehistoric Mammals of Australia and New Guinea: A hundred million years of evolution* by Long J, Archer M, Flannery T and Hand S (2002).

Family	Genus	Early Cretaceous 120–100mya	Paleocene 63–61mya	Late Oligocene 26–23.5mya	Early–Mid Miocene 23.5–11mya	Pliocene 5.3–1.75mya	Pleistocene 1.75mya–10 000	Holocene 10 000–0
Kollikodontidae	***Kollikodon****	**1**						
Steropodontidae	***Steropodon***	**1**						
	Teinolophos	**1**						
Tachyglossidae	***Megalibgwillia***				**1**		**2**	
	Zaglossus						1	3
	'Z' hacketti						**1**	
	Tachyglossus						1	1
Ornithorhynchidae	***Monotrematum***		**1**					
	Obdurodon			**2**	**1**			
	Ornithorhynchus					1	1	1

Monotreme ancestry

A team of dirty and bedraggled palaeontologists working in adverse weather conditions using explosives, front-end loaders, picks and shovels, and even tiny paint brushes to separate lots of earth or rock from between tiny fossils, is starkly contrasted with white-coated molecular biologists working in a spotless high-tech laboratory. However, information from both of these areas of research is now being used in an attempt to discover what was happening on the Earth millions of years ago. The key, of course, is the molecule DNA (deoxyribonucleic acid) which carries the information for almost all biological processes, but which also changes through mutation, producing variations in the structure and function of organisms which are the basis of evolution. Very old fossils of course do not yield DNA that can be analysed. However, genetic changes can be deduced from the analysis of the DNA of living organisms. Sophisticated methods used by molecular biologists, but incomprehensible to mere mortals, and with names like 'mitochondrial protein-coding gene analysis', 'DNA–DNA hybridisation' and 'nuclear gene cloning', have been used in an attempt to determine the date of the platypus and echidna separation. Although the estimates vary depending on the assumptions used in each of the methods, it seems to be agreed that the echidnas and platypus ancestral lines parted from each other relatively recently (65 to 23 million years ago) and well after the specialised platypus-like fossil forms came into existence. Unfortunately, this does not support the view held by many palaeontologists.

The oldest echidna fossil so far discovered is only 14 million years old, suggesting this group of monotremes probably evolved after the platypus-like forms, known to have been scattered throughout the past 100 million years of fossil history. It seems unlikely that the terrestrial, toothless, narrow-snouted, robust, sticky-tongued and spiny forms of monotreme like the echidna, would have evolved from a broad-billed, dense-furred, aquatic or amphibious platypus-like ancestor. Most palaeontologists consider that both groups are so specialised that they must have arisen from a much earlier, more general ancestor. Given the difficulties of investigating events that happened a very, very long time ago, the present understanding of the evolutionary history of the platypus and its relatives is perhaps part of a jigsaw puzzle from which most pieces are still missing.

Although known to have had several fossil relatives, *Ornithorhynchus anatinus* is the only living species of platypus. It is indeed a very specialised animal. As such, an environmental change might quickly drive it to extinction, just as its relatives succumbed to such changes through geological time. Yet, as a water-dependent species on a very dry continent, the modern platypus has survived for at least five million years and has so far weathered the dramatic ecological changes wrought since Europeans occupied the Australian continent just over 200 years ago (Chapter 8). Modern humans (*Homo sapiens*) on the other hand – all of whose ancestors are also extinct – have been in existence for only about 200 000 years and counting.

8 CONSERVATION: PLATYPUSES AND PEOPLE

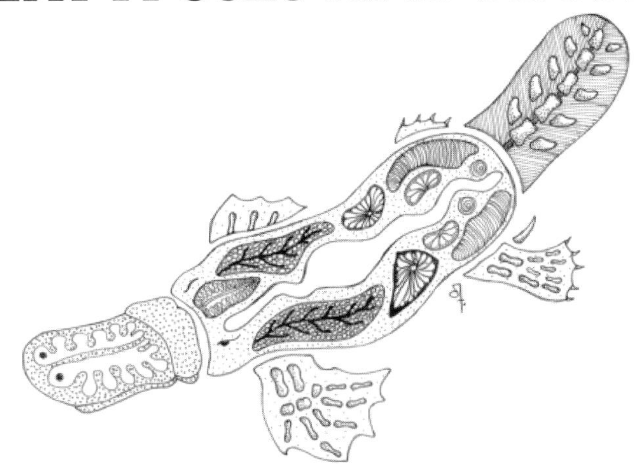

'The natives sit upon the banks, with small wooden spears, and watch them everytime they come to the surface, till they get a proper opportunity of striking them. This they do with much dexterity; and frequently succeed in catching them in this way'.

'Governor Hunter saw a native watch one for above an hour before he attempted to spear it, which he did through the neck and foreleg; when on shore, it used its claws with so much force, that they were obliged to confine it between two pieces of board, while they were cutting off the barbs of the spear, to disengage it'.

SIR EVERARD HOME (1802)

Relating Governor John Hunter's 1797 platypus observation at a lagoon adjacent to the Hawkesbury River near Richmond in New South Wales.

The preserved skin of the unlucky animal, which had been speared and opportunistically observed by Governor John Hunter in November 1797, was transported to England from the colony of New South Wales. It initiated controversies between leading biologists from Britain, France and

Germany which lasted for decades. There was argument over the correct naming and if it was in fact a mammal that suckled its young. The greatest area of contention, not resolved until 1884, was whether it gave birth to live young or laid eggs. It is now known, of course, that *Ornithorhynchus anatinus* is indeed a mammal that suckles its young on milk and lays eggs.

Governor Hunter's drawing of the platypus, described as 'ghastly' by Merv Griffiths in his 1978 book *The Biology of the Monotremes*, is a very poor representation of the living animal (Figure 8.1). However, this illustration, the preserved skin and Hunter's own description of this 'small amphibious animal of the mole kind' appear to represent the first awareness of the existence of the platypus by Europeans, or more particularly, by European science. Other non-indigenous visitors to Australia probably had already seen the strange animal prior to Hunter's report and the indigenous people, of course, had known of its existence for thousands of years before Europeans first arrived in Australia.

ORNITHORHYNCUS PARADOXUS.

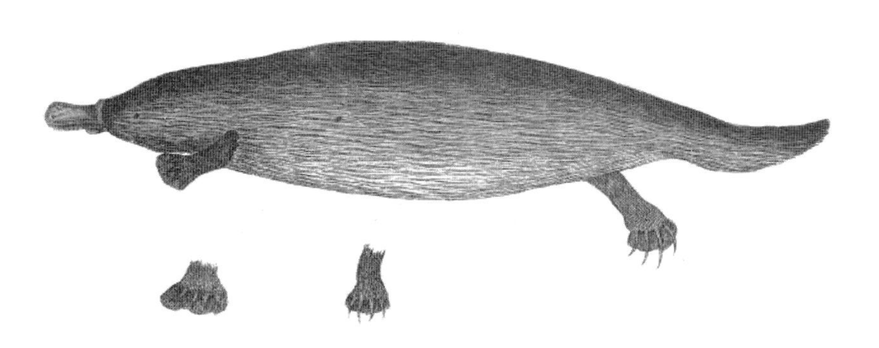

AN AMPHIBIOUS ANIMAL of the MOLE KIND,
which Inhabits the Banks of the fresh water Lagoons in New South Wales — its fore feet are evidently their principal assistance in Swimming & their hind feet having the Claws extending beyond the Webb'd part are useful in burrowing.

Figure 8.1 Illustration of the platypus drawn by Governor John Hunter, published by Collins, 1802. Photo: State Library of Victoria

The platypus and Aboriginal people

A considerable variety of different languages, several of which have fortunately survived, were spoken by groups of Aboriginal people that occupied mainland Australia and Tasmania in the past. Searches of the historical literature concerning Aboriginal people, their language and culture yielded valuable information about the possible pre-European distribution of the platypus. A range of recorded names given to the platypus was found. For example, the word 'mallangong' was apparently used by peoples from the Goulburn and Bathurst districts of New South Wales and 'mallunggang' was the name given to the animal in the area around Bairnsdale in Victoria. The name 'tambreet' appears to have been used in the Yass and Tumut areas, quite close to the Goulburn region. West of Bathurst, along the Macquarie and Castlereagh Rivers and their tributary streams, around Dubbo and Mudgee, variations of the word 'billadurong' were given to the platypus. In Tasmania most names for the platypus were found to be very different from any used on the mainland, and those applied to the animal by Aboriginal people from Queensland were again often very unlike the names used in areas of south-eastern Australia. The occurrence of names for the platypus, mention of their use as food or for furs and some references to their cultural significance seemed to indicate that the current overall distribution of the species incorporates most areas where it occurred in pre-European times.

The early naturalist George Bennett was told by his Aboriginal helpers from the Yass region of New South Wales that the platypus was 'very good to eat' and the early literature certainly records it being included in the diet of Aboriginal people living along the coastal and tableland streams of New South Wales and Victoria. Its use as a food source, however, appears to have been less common along the streams flowing west of the Great Dividing Range in these states. The Aboriginal peoples of Tasmania also consumed platypuses, as its distinctive leg bones and cranial material have been found in the archaeological excavation of several caves in south-western Tasmania, which were used by indigenous people during the later Pleistocene period (16–30 000 years ago). The most numerous prey species hunted at that time appeared to have been Bennett's wallabies (*Macropus rufogriseus*), which today are very lean animals, with a low proportion of fat in their carcass composition. Archaeologists have suggested that platypuses could have been selectively taken as a valuable source of fat in the diet of Tasmanian Aboriginal people during this period, when temperatures were estimated to have been 5°C cooler than today.

The naturalist Richard Semon, working on the Burnett River in northern Queensland during 1892/93, reported his Aboriginal workers readily being prevailed upon to collect echidnas, as this species was considered by them to be a favoured food item. They were, however, less keen to seek out platypuses which they apparently were not enthusiastic about eating. In spite of this report, the presence of platypus skeletal material in an Aboriginal rock shelter excavated in southern Queensland indicates that the platypus was also included in the diet of indigenous people, at least in parts of Queensland.

The platypus and Europeans

Soon after their arrival, Europeans began to interact with the platypus. There was keen scientific interest in the animal, but it was also killed for other purposes. In 1815, during a trip across the Blue Mountains to the now Central West of New South Wales, Governor Lachlan Macquarie recorded in his journal that one of his companions, Sir John Jamison, returned to camp with 'a large string of fine fish and 2 water moles [platypuses].' Such restricted hunting of wildlife for food or for recreation by Europeans, or use of it as a resource by indigenous peoples, would probably have had minimal impact on most wildlife species in Australia, including the platypus. On the other hand, commercial hunting, fishing, forestry, agriculture and other economic pursuits, which progressed from the first occupation of Australia by European settlers, were to have much greater effects on the continent's wildlife. Very soon after settlement began, changes in the environment of Australian flora and fauna were recorded. For example, the botanist George Cayley, travelling in a rural area of the colony of New South Wales in 1802, documented the erosion of banks and deposition of sediment in the river system, and attributed this to the unrestricted presence of cattle. In 1803 a General Order from Governor King stated that 'from the improvident method taken by the first settlers on the sides of the Hawkesbury [River] and creeks in cutting down timber and cultivating the banks, many acres of grounds have been removed'. Fortunately, it appears that these activities have affected the platypus much less than many other vertebrate species in Australia. The platypus, however, has not been totally exempt from the impacts of humans on the environment.

Hunting

Platypus fur is very fine and dense, the pelage is short and has beautiful colour and texture. Although highly insulative, a platypus skin is quite small and many would have been needed for a garment. This, however, did not appear to deter European hunters, who slaughtered thousands of platypuses for their pelts. To avoid damaging the coat, hunters stunned the animals by firing a heavy calibre shot in the water underneath them. A 'gun dog' or retriever then swam out to fetch the stunned platypus. This historical practice is almost certainly the source of the anecdote, often found even in current articles and books, that the venom of the platypus is not fatal to humans, but can kill a dog. Unsurprisingly, dogs are not at risk of being chased and spurred by marauding platypuses in the wild. Historically it is more likely that a male platypus, recovering from being stunned by a hunter's bullet, sometimes spurred the retrieving dog in the muzzle. This undoubtedly caused the dog excruciating pain and in certain instances may have resulted in death. The effects of the venom on humans, also causing great pain but not death, were discussed in Chapter 3.

Because of the thickness of its skin and the resultant stiffness after tanning, pelts of the platypus were apparently not in great demand for clothing or for export but many were used locally to make hats, slippers and rugs (see Colour Plate (bottom), page 36). As a single rug required 40 to 60 pelts, it is clear that large numbers of platypuses must have been killed during the period in which they were hunted commercially. An article in the *Victorian Naturalist* journal noted that around 1870 a local resident of a Melbourne suburb made 'a fairly good living' by selling platypus skins to local furriers. The naturalist Harry Burrell, in his 1927 book *The Platypus*, recorded that 'one of two brothers who were both great hunters of platypus, confesses to having been wicked enough to have shot many thousands during his thirty-two years of work'. These depredations occurred up to, and apparently for some years after, the time the species was protected by legislation in the various states in which it was present. In 1892 the platypus was declared a protected species in Victoria, with New South Wales, Queensland and Tasmania following between 1901 and 1907. Disturbingly, the species was not fully protected in the state of South Australia until 1912, by which time it was probably extinct from the Mount Lofty Ranges and Adelaide Hills. In New South Wales, the sale of any platypus furs was banned in 1919 in an attempt to eliminate illegal hunting. A fisheries biologist, who carried out native fish surveys in the

Murray and Murrumbidgee Rivers during the late 1940s, reported the son of a grazier from a property near Darlington Point on the Murrumbidgee River sleeping under a platypus rug on the veranda of the homestead.

Freshwater fishing

Early inland (freshwater) commercial fishing in New South Wales almost certainly caused significant platypus mortality, owing mainly to the use of nets with small mesh sizes, but also because the fishery included most of the state's rivers, except for the upland streams, where trout and other species of salmonid fishes had been introduced for angling and were protected by the *Fisheries Act* in 1902.

By-catch mortality of air-breathing vertebrates, including the platypus, diving birds and turtles, has been recognised as a significant risk in all states where platypuses occur. This has resulted in a range of regulations and modifications to fishing gear aimed at reducing by-catch of non-target species, although little direct research or monitoring have been done to assess the effectiveness of these regulations and initiatives. One study in New South Wales found that platypuses could easily negotiate the 9-cm mesh, previously set as a by-catch reduction measure for a variety of fish traps, including those used to capture various species of freshwater crayfishes (yabbies). Since these experiments, yabby traps have been banned from streams within the current distribution of the platypus in Tasmania, Victoria and New South Wales. Only in Queensland are the most common type of yabby traps, the 'opera house' traps (so called, because of their assembled shape resembling the Sydney Opera House) still legally used in streams where platypuses are found. By-catch mortality has been reported from Queensland as a result, and continues in other states where many people are unaware of the effects and/or the illegality of using these traps.

The drowning of several platypuses in a single yabby trap has been reported on a number of occasions although the attraction of platypuses into these traps is not fully understood. However, platypuses are known to locate their prey by sensing the electric fields and water movements generated by prey species (especially large food items such as yabbies; Chapter 4), and this may result in their being lured into traps that have already captured yabbies. Platypuses are also drowned in other illegally set fish traps, 'gill' (mesh) nets and set lines (see Colour Plate (top), page 37). Nets are often used by researchers to capture platypuses in pools but these nets are watched constantly and no weights are placed on the bottom of the nets. Any entangled platypuses can thus easily bring the net to the surface (Figure 6.1).

Fortunately, commercial fisheries in inland waters are now quite limited in all states where platypuses are found. No commercial or recreational net or trap fishery for native fin fish species now exists in Queensland, New South Wales, Victoria or Tasmania, mainly due to the dramatic decline in the abundance of these species brought about by the intensity of the early fisheries and as a result of habitat change. However, there are still small commercial eel fisheries, based on the use of fyke nets (Chapter 6) in limited numbers and sections of streams in Victoria and Tasmania. In Queensland, eels are commercially captured in baited traps only in farm dams on private properties and in a few impoundments on public waterways. Commercial capture of the two species of eels (short-finned, *Anguilla australis* and long-finned, *A. reinhardtii*) in New South Wales is restricted to estuaries, farm dams and a few artificial water impoundments. Introduced Common or European carp (*Cyprinus carpio*) are now targeted in New South Wales and Victoria in limited commercial fisheries using a variety of gear, including gill nets and electrofishing equipment.

Occasionally anglers will capture a platypus on a worm, spinner, lure or even a fly. How frequently this occurs, and its possible consequences to local populations, is not known (see Colour Plate (bottom), page 37). Platypuses caught in this way should not, of course, be released with the hook still embedded in the bill or body, as this is likely to result in death from starvation or infection. A male platypus can be handled safely by the tail and it is possible to wrap it up in a Hessian bag, jacket or some other bulky garment to immobilise the back legs (and spurs), while the hook is removed. Discarded fishing line can also entangle wildlife, including the platypus, and result in injury or death.

Organic pollution

Ever since European settlement, first on farms, villages, then towns and cities, the effluent produced by humans and their domestic stock has resulted in water supplies becoming unusable for drinking and sometimes even for stock-watering. It might have been expected that this disturbance to the natural environment would have affected the platypus. Sewage effluent entering streams most often leads to a decrease in the diversity of macroinvertebrate species, with certain tolerant ones becoming more numerically dominant in the benthic fauna and more susceptible ones declining in their distribution and/or abundance. However, although platypuses can sometimes be selective in their choice of prey species, they mainly consume these organisms from the bottoms of the creeks and rivers in relation to their abundance or biomass, rather than their diversity (Chapter 5). As a result, it appears that non-toxic organic pollution from farms, meatworks, dairy factories and sewage treatment plants has not had a markedly detrimental effect on platypuses. Individuals have been observed foraging in streams unsuitable for human primary contact (swimming) within the zone of influence from country sewage treatment plants discharging secondary or tertiary treated effluent. Platypuses have been found in streams affected by septic tank leakage and overflows. Although there have been no studies directly investigating this, current knowledge indicates no effects in terms of clinical toxicity or disease in the species arising from exposure to treated human and animal sewage effluent. A platypus was even observed swimming and apparently foraging in a sewage settlement pond in a treatment plant in northern Tasmania. Thus, the idea that the platypus could be considered a biological indicator of clean water is quite unrealistic!

Dams, weirs and other structures in streams

As farms and settlements spread away from the centres of each newly established colony in Australia, streams were dammed for human water supply, for stock and finally as a source of irrigation water for crops. Weirs were built to control river levels, and although usually less than 5 metres high they, like dams, presented a barrier to the movement of native fishes and platypuses. Stream crossings or culverts were established on roads between settlements and these often created small- to medium-sized ponds upstream and barriers downstream. When flying over any country area in Australia, a mosaic of lakes, ponds and small pools extends across the landscape. Almost all of these result from the modification of waterways by humans.

If entrances to the pipes or channels of culverts are more than 20 cm vertically above the stream, platypuses can be prevented from accessing and using the culverts. This forces the animals to move overland, increasing the danger of predation or being killed by vehicles, especially where the culverts are under main roads. Research indicates that platypuses regularly use culverts but there are no accurate estimates of flow velocities which could prevent them from swimming against a current within such narrow structures. Swimming against a stream velocity of

1–1.2 metres per second has, however, been observed in one individual moving through an artificial fishway and in another within a natural stream riffle, indicating that platypuses could move through culverts where such flows occur. In Tasmania, platypuses are quite often seen moving over land. This is thought to be related to the fact that, until recently, there have been no introduced foxes on the island and so the risk of predation has been less than on the mainland. Whether this suggestion is an accurate one has not been tested but unfortunately the apparent penchant of Tasmanian platypuses for leaving their streams and taking to the land results in roadkills. The road sign shown in Figure 8.2 highlights this phenomenon, which is fortunately quite rare on mainland Australia.

The downstream effects of river regulation include temperature, flow and sediment changes, which often result in altered benthic communities and reduced foraging areas below large dams. Lowered water temperatures, characteristic of water released from below the thermocline in most impoundments in Australia, place an energetic demand on platypuses living downstream from large dams. The species is physiologically well adapted to living under cold conditions in winter over much of its current distribution. But raised metabolic demand, coupled with changes to macroinvertebrate food availability, must impose additional stress on animals foraging in waters downstream of large dams. Most foraging by radio-tracked platypuses during operational releases for irrigation from Eildon Weir on the Goulburn River in Victoria was observed to take place in slower-flowing backwaters during higher flows, an observation which corresponds well with the findings of research on the energetics of foraging and diving (Chapter 5).

Upstream increases in water levels, associated with dam construction and operation, change relatively shallow and productive flowing stream and river environments into deep, less productive lake systems. Platypuses appear to be unable to forage successfully for small food items at depths greater than about 5–10 metres and are therefore only occasionally reported in deep areas of water storage impoundments. For example, only four records of platypuses from 32 deep (>10 metres) impoundments in New South Wales were recorded from a survey of platypus distribution in the late 1980s, while 30 records in the survey came from the shallower headwaters of these storages. Smaller pondages behind weirs less than 5 metres in depth, however, are often occupied as part of the foraging range of platypuses and may represent breeding areas if banks are suitable.

During low stream flows, such weir pools often represent significant refuges for resident platypus populations in drought affected areas.

Planning and building of large dams over the past two decades has been more common in Queensland than in the other states, where numerous large dams already exist. More consideration is now being given to conditions downstream from dams – including the implementation of operational procedures for new and existing structures to reduce their downstream impact – by providing environmental flows, simulating natural flood events, reducing cold water pollution and increasing oxygen levels of the released water. However, the success of these measures in maintaining the integrity of ecosystems downstream of large dams, including platypus habitat, are as yet poorly researched and understood.

Figure 8.2 Platypus traffic sign in northern Tasmania, where platypuses have been seen or killed on this road. Photo: Peter Temple-Smith

Unfortunately the monitoring and adaptive management programs, which normally arise from environmental impact assessments, only permit relatively small mitigating adjustments to be made. Should a severe impact be highlighted by such monitoring, the established infrastructure normally makes it impossible to reverse any such serious effects of a project. For that reason it is essential to assess fully the possible impacts well before projects proceed. Within various jurisdictions – state, local and federal – many different legislative instruments (e.g. acts, regulations, plans and strategies) have been used with varying success to protect the environment of the platypus and other wildlife species.

Salinity

In Australia, increasing salinity of inland waters is seen as a major threat to the health of aquatic ecosystems and water resources in parts of many catchments. No studies investigating the effects of salinity on either the foraging behaviour or osmoregulatory physiology of the platypus have been published. However, platypuses were recorded close to the sampling points on seven streams in New South Wales where the water quality was considered only 'fair' with regards to salinity (0.34–1.02 parts per thousand of salt; conductivity of 500–1500 microsiemen units per centimetre). This suggests that the species can tolerate the levels of salinity currently found in a number of streams within their current distribution. Apparently the electrosense of the platypus functions less efficiently as the conductivity of water increases, leading to a suggestion that the reported absence of the platypus from many inland streams could be related to its foraging being less effective in those with elevated salinities. However, there is little support for this hypothesis as platypuses are absent from apparently suitable inland streams unaffected by salinity, and can occur in saline streams, in upper estuary situations and occasionally in more saline water.

Introduced species

Many people who have observed platypuses in the wild mention willows (*Salix* spp.) in their description of the habitat in the area in which they made their observations. A study in the upper Shoalhaven River in New South Wales also found that 76% of the burrows used by radio-tracked platypuses were among the roots of one or more willow trees. Another radio-tracking study in a small stream in Victoria, on the other hand, showed that platypus foraging activity was positively correlated to the presence of native vegetation and introduced poplar trees, but negatively correlated to the presence of willows.

Species and varieties of the willow genus *Salix* in Australia are reported to be involved in adverse effects in streams, including increased local flooding, reduced summer stream flows, restricted fish passage, loss of habitat for aquatic macroinvertebrates and low dissolved oxygen. These effects appear to be most pronounced in small streams, particularly where willows occur in high densities, often forming dense root mats which platypuses cannot penetrate to dig their burrows (Figure 8.3). The management of willows is currently the subject of considerable debate. In one reported instance where willows were removed and replaced by native species of vegetation, resident radio-tracked platypuses exhibited no apparent adverse effects from the river management activities. Similarly, in another study, no difference was found in the numbers of platypuses observed at a study site on the Wingecarribee River in New South Wales before and after willows had been stripped from about 1.75 km of the stream. River management literature normally proposes that where willow removal is considered essential, their removal should be planned and

Figure 8.3 Willow regrowth can choke small streams and dense root masses can make the banks unsuitable for burrowing by platypuses. However, established adult willows can have an important role in consolidating stream banks.

executed very carefully, native plantings should be established before the willows are completely removed and the frequent proliferation of weed species (as a result of increased light availability) needs to be diligently controlled. Research shows that increased light and temperature from changes to riparian vegetation can result in dramatic alteration of the distribution and abundance of aquatic invertebrate species and in-stream productivity. Currently there are no published studies that quantitatively investigate the effects of willow tree removal on platypus populations, their habitat or the availability of their food.

Common carp (*Cyprinus carpio*) were probably first introduced into Australia around 1850 but did not spread until the introduction of a particular strain in the 1960s. This strain was used in fish farms but was transferred quickly into streams in the wild. Ecological effects of high densities of carp are poorly understood, but increased bank damage, disturbance of aquatic plants and turbidity are all reported consequences. Food competition between carp and the platypus is also possible. The overall disruption of riverine food webs by the huge biomass of carp found in certain aquatic systems may be very complex but are presumed to be detrimental to the freshwater systems in which the species occurs. However, the current distributions of platypuses and carp certainly overlap in many places. For example, they are found together in numerous stream systems of the Murray–Darling Basin, as well as in many water bodies of coastal Queensland, New South Wales and Victoria. The species is also present in restricted areas of Tasmania.

Two species of introduced salmonid fishes, the brown trout (*Salmo trutta*) and rainbow trout (*Oncorhynchus mykiss*), also compete to some extent with the platypus for food. It has been suggested that the trout may be having a deleterious effect on platypuses, but these species also frequently overlap the platypus in their distribution and have done so for over 100 years. Several other introduced species of fish also overlap parts of the platypus distribution. These include redfin perch (*Perca fluviatilis*) in parts of Tasmania, Victoria, New South Wales and Queensland, plague minnows (or mosquito fish; *Gambusia holbrooki*) in most states and the Mozambique cichlid (or Tilapia; *Oreochromis mossambicus*) in northern Queensland. All have some overlap in their prey species but nothing is known of any effects of this competition on the platypus.

Intuitively, the introduction of one or more competing feral fish species into a riverine food web, including a breeding population of platypuses, would be expected to have an adverse impact, especially when carp and/or salmonids often constitute a very large part of the total vertebrate biomass in certain streams at specific times of the year (e.g. winter spawning runs in the trout species). Although not well investigated by studies, there are currently neither data indicating poorer condition of platypuses in areas where their distribution overlaps with one or more introduced fish species, nor conclusive evidence of changes in distribution or abundance of the platypus as a result of these fish species being present. An observation made during a platypus study in the Murrumbidgee River, near Bredbo in New South Wales in 1970, even suggested that platypuses were more successful than trout at acquiring food. During summer and autumn especially, captured platypuses were found to be in good physical condition but trout caught at the same time appeared to be in poor condition.

Disease

Platypuses are known to carry a number of parasitic animals in the wild, including their own unique species of tick, *Ixodes ornithorhynchi*. The early stages (instars) of this species are found, often in high numbers, around the rear ankles and back legs of animals while larger adults occur in smaller numbers on the front legs and in the body fur (see Colour Plates page 35). The tick is thought to transmit the protozoan blood parasite *Theileria ornithorhynchi* found in the blood of most platypuses examined, but the parasite is usually regarded as innocuous, normally infecting only a low percentage of red blood cells. Blood tests done on 131 platypuses from the upper Shoalhaven River in New South Wales found that 50% had antibodies to the bacterial infection leptospirosis, indicating they had been exposed to the disease. However, no clinical symptoms of the disease were found. The type of leptospirosis identified (*Leptospira interogans*, serovar *hardjo*) is probably transmitted by infected cattle urinating in streams. Platypuses are also subject to a number of other viral, protozoan and bacterial infections but little is known of their prevalence or effects.

Platypuses suffer from a fungal infection (*Mucor amphibiorum*) found in amphibians on mainland Australia. This disease was possibly transmitted to Tasmania by green tree frogs (*Litoria caerulea*) unintentionally hitching a ride on bunches of bananas. To date, the infection has only tentatively been reported in a few mainland platypuses but in parts of Tasmania it has resulted in a condition that causes severe skin ulcers (Figure 8.4). It

Figure 8.4 A skin ulcer on a Tasmanian platypus caused by an infectious fungus, *Mucor amphibiorum*. Photo: Joanne Connolly

can invade other tissues, including the lungs, and has led to mortalities in certain populations. The strain of this fungus found in Tasmanian platypuses appears to be more virulent than that found in amphibians on the mainland, which may be the reason for the infection and deaths in that state. This disease has been reported in other parts of Tasmania but was first and most often recorded from streams of the Tamar River system in the north-east of the state.

Platypuses on mainland Australia possibly evolved with the fungus, while those on the island of Tasmania – separated from Australia for at least 12 000 years – may have only recently become exposed and so are less resistant to infection. Studies have shown healing of ulcers and apparent recovery from the disease in some individuals, suggesting development of resistance may be possible in infected populations. It has also been suggested that persistent organic pollutants, such as pesticides and PCBs (polychlorinated biphenyls), finding their way into streams, might be causing immune system suppression in the platypuses of Tasmania. Although elevated levels of some of these persistent organic pollutants (DDT and its metabolites, Lindane and PCBs) have been found in platypuses in Tasmania, they were not at levels which have caused immune suppression in other species.

Urban development

Platypuses are often reported close to fairly large rural population centres, including Canberra. They also still occur within parts of the metropolitan areas of Hobart, Melbourne, Sydney and Brisbane. Nevertheless, it appears that both the numbers and distribution of platypuses in these areas have been significantly reduced over historical time. The naturalist George Bennett, for example, used to collect platypuses from the streams of north-western Sydney in the late 1850s. There are still occasional reports from the outer suburbs of Sydney today, but the platypus is now considered very uncommon or even extinct in these badly degraded streams. The observed reduction in occurrence or numbers has been attributed to the various effects of urban development. These include pollution, sediment accumulation, bank erosion and increased flash flooding associated with rapid run-off from impervious surfaces, like roads, roofs and parking areas. In a recent study carried out in catchments around Melbourne, no platypuses were found in streams with extensive impervious areas in their catchments. Very few were found in streams where sediment run-off was high and there were elevated levels of nitrogen, phosphorous and several heavy metals. In these urban streams around Melbourne, and in those found in country towns and residential centres around Australia, a host of items finding their way into streams, by way of stormwater drains, have resulted in platypus injury and death (Figure 8.5). The Australian Platypus Conservancy, for example, studied six streams close to Melbourne, and found that 14 out of 133 platypuses captured (i.e. more than 10%) were entangled by items such as fishing line, elastic bands, sealing rings from food jars and even an engine gasket.

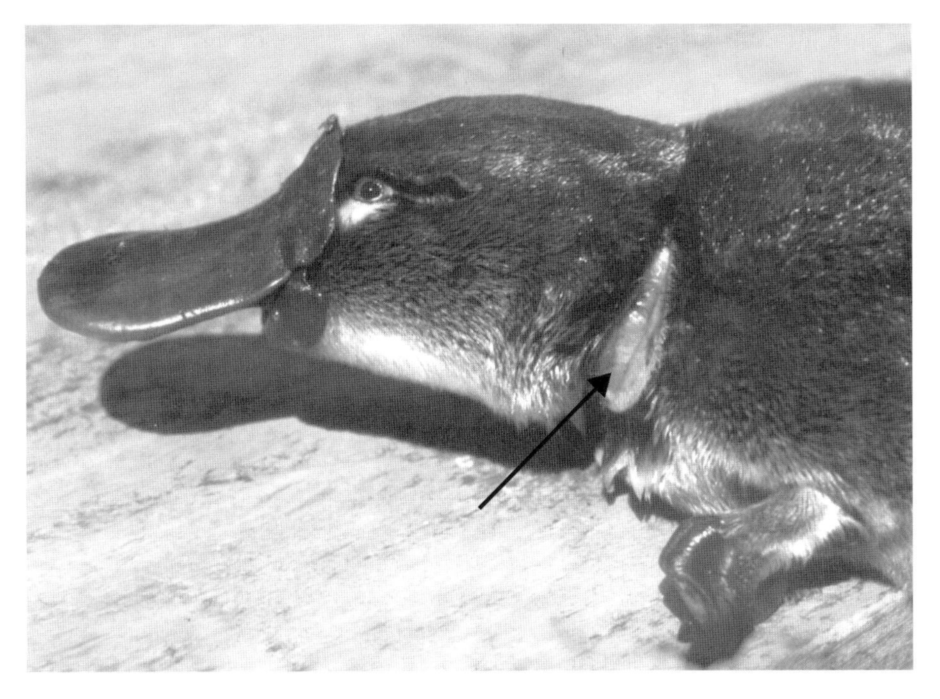

Figure 8.5 The deep wound in the neck of this platypus was inflicted by a plastic wrist band entangled around its neck and front foot, which was also injured. Fortunately, in this instance, a veterinarian was able to suture the wounds with dissolving stitches and the platypus was released. Photo: Peter Tonelli

Forestry

Platypuses are frequently recorded in forested areas, including those subjected to logging and associated activities. A survey of platypus distribution funded by the New South Wales National Parks and Wildlife Service in the late 1980s reported platypus sightings in State Forests in New South Wales within every type of forest management at that time – from reserves to selective logging and clear felling. A study around the town of Bombala on the southern tablelands of New South Wales also recorded the presence of platypuses in most streams in the area, despite logging in their headwaters for over 100 years.

The main environmental effects attributed to forestry operations which are likely to affect the platypus are those that result in damage to stream banks and reduction in the abundance and diversity of benthic macroinvertebrate species in streams. The effects of poorly managed forestry operations can include elevated stream flows, filling of interstitial spaces in stream substrates, sedimentation, elevated suspended solids, bank damage as a result of the removal or destruction of riparian vegetation and the use of pesticides and herbicides. In a study carried out during 2000–2001 in previously clearfelled (15 years prior to the study) and undisturbed forests in the South Esk River catchment of north-eastern Tasmania, fewer platypuses were captured in the streams of the previously cleared forest. In addition, more first-order streams (first streams in the catchment) in the logged forest were found to be devoid of platypuses when compared to streams in the undisturbed areas. The authors of this study attributed these findings to the adverse impact the previous forestry operations had on the quality of platypus habitat, including changes in bank stability, overhanging riparian vegetation, suitable foraging substrates and food availability.

Agriculture

Dramatic changes to the Australian landscape occurred soon after European settlement began but gathered pace with the advent of mechanised farming and its associated demand for water. Land clearing, surface run-off from farms, gully and stream bank erosion (Figure 8.6) and the damming of streams for water supply resulted in changes to stream flows, sedimentation and turbidity (Figure 8.7), salinity problems in catchments and the influx of chemicals, including nutrients and pesticides. These changes in the physical environment quickly manifested themselves as changes in the diversity and abundance of aquatic species. Indigenous aquatic species often declined, both in variety and population numbers, while feral plants and animals flourished as a consequence of habitats more suited to their survival and reproduction.

Platypuses are frequently reported in agricultural areas. Three separate surveys in New South Wales recorded 52–76% of platypus sightings on farmland. It would, however, be a mistake to be complacent about these observations and to consider agricultural use of the land as benign with respect to platypus conservation. Despite the apparently common occurrence of platypuses in agricultural areas, there are examples of reduction in numbers and/or fragmentation of distribution occurring in streams of the Eden area and in the Bega, Thredbo and Richmond Rivers of New South Wales. Geomorphological changes in the channel of the Bega River have resulted in sand accumulation and stream bank degradation. These changes are considered irreversible and have been attributed to early

Figure 8.6 Excavation of a site for this irrigation pump on the Hastings River resulted in stream bank erosion. Oil or fuel leaking into a river from pumps can mat the feathers of water birds and the fur of platypuses and water rats, impairing their thermal insulation.

Figure 8.7 Sediment draining into the Hastings River after heavy rain in the headwaters of one of its tributaries.

agriculture practices in the area. However, a study in the nearby Bombala area showed that platypuses occurred in most streams, in spite of this area having been used for both cattle and sheep grazing for approximately 160 years. Data collected by the Australian Platypus Conservancy in Victoria showed fewer than 200 individuals now occupy the Wimmera–Avon River system (a large catchment of over 2 400 000 hectares) and the species has apparently recently become extinct in the neighbouring Avoca River basin. In each of these instances the changes have been mainly attributed to poorly managed agricultural practices, including unlimited access of stock to the streams (Figure 8.8), water abstraction, removal of indigenous riparian vegetation, followed by a resultant accumulation of sand and the reduction in pool-riffle sequences found in streams.

Fortunately, many government and community programs and initiatives are now attempting to control catchment erosion and to rehabilitate streams and riparian zones in rural areas, but some geomorphologists suggest that the health of a number of streams in Australia will remain permanently impaired as a result of these accumulated human-induced changes.

Little is known of the effects of agricultural chemicals in the streams inhabited by platypuses. Persistent organochlorine residues of the agricultural pesticides DDT and Lindane were found in the fatty tissues of all the platypuses in a Tasmanian study and in certain specimens sampled on the mainland. However, the levels reported did not appear to have had any clinical effects on the platypuses or to be at levels which could affect their breeding.

The future of the platypus?

Over a period of only 20 years, deep pools with a mixture of sand, woody debris and cobble/gravel substrate, in a series of pool-riffle sequences in the upper Shoalhaven River area in New South Wales, have changed into small, shallow pools linked by meandering sandy channels (see Colour Plates, pages 38 and 39). The disappearance of deep pools has been of great concern, not only because of the assumed reduction in productivity of macroinvertebrate prey species, but also the loss of platypus refuge pools during periods of extended drought. During the severe drought of 1979–1983 the upper Shoalhaven River ceased to flow (Figure 6.6), but platypuses continued to occupy and breed in the large deep pools

Figure 8.8. Removal of riparian vegetation, unrestricted cattle access, wombat activity and intermittent floods over a period of approximately 20 years dramatically changed a section of the upper Shoalhaven River from a deep narrow pool in which platypuses were often captured (top), to a broad and shallow one where platypuses are now only occasionally seen (bottom).

which remained. The decline in the number of refuge pools during future extended droughts was expected to impact more significantly on the resident platypus population. However, there was no evidence of significant loss of body condition in individuals captured during the short but relatively severe drought of 1994–95. Up until the 2006/07 breeding season during the most recent extended drought beginning in 2001, there appeared to be little impact on body condition or recruitment of young platypuses to the population in the upper Shoalhaven River.

The effects of past and present poor land management practices appear to represent the most immediate potential threat to platypus populations, and yet individuals in the upper Shoalhaven River population seem little affected by rather dramatic changes in their habitat. Clearly more research effort is required to understand the reasons for such apparent anomalies. However, there is serious concern that, although many populations appear to be largely unaffected at present, they could exhibit rapid and severe changes in response to such degrading processes in the long term. The fragmentation and extinctions found in parts of certain river systems suggest the possibility that apparently secure local platypus populations could quickly become threatened or extinct as a result of change arising from human activities.

Currently the International Union for the Conservation of Nature and Natural Resources (IUCN) categorises the platypus as a species of 'least concern'. This is largely because, with the exception of their historical presence in the Mount Lofty Ranges and Adelaide Hills, the platypus has continued to occupy aquatic ecosystems throughout most of its historical range. It continues to occupy and breed in ecosystems where various threatening processes are having major impacts on aquatic and riparian habitats and on the distribution and abundance of other species, particularly native fishes.

The future conservation of the platypus will depend on halting or reversing all the human impacts discussed in this chapter, especially those resulting from poor land management practices and the development and use of water resources. As a result of the progression of global warming, Australia will inevitably experience a new climate regime which will also impact severely on water-dependent species, including the platypus. The competition between humans and these water-dependent species will intensify as water resources decline or are redistributed. Platypuses now appear secure (or of 'least concern') over much of their distribution. With changing climate, it remains to be seen which areas of this distribution will continue to receive sufficient rainfall to maintain these populations. The platypus is an integral part of many freshwater ecosystems in south-eastern Australia and is considered a national icon. Its unique evolutionary position and relationships also make it a species of global importance. In their book entitled *Conservation Biology in Theory and Practice*, the late and influential Australian ecologist, Graeme Caughley and his co-author Anne Gunn indicate that 'the signal of a conservation *problem* is the sustained decline rather than the final stage of low numbers [of a species]'. The currently common occurrence of the platypus throughout most of its historical range is not a reason for complacency. Targeted research, planning, informed decision-making and careful management are all required to ensure that 'the animal of all time' does not slide into becoming a conservation *problem*, but instead continues to be a prime example of Australia's unique fauna.

9
QUESTIONS, ANSWERS AND MISCONCEPTIONS

Walkabout magazine (1 May 1948) with permission from Foster's Group Limited

In response to the question 'have you ever seen a platypus in the wild?' most Australians or travellers to Australia would probably answer 'no'. In spite of this, however, the existence of this unique animal is well known. The platypus is an Australian icon species about which many people care a great deal, but it is a species about which people often know very little or have misconceptions. As highlighted in other chapters of the book, many controversies have been associated with the platypus and a number of dubious 'facts' are accepted and frequently quoted. In this chapter common queries and various myths are presented as a series of questions and answers. In some instances no definitive answer can be presented, as the science is poorly known or understood. At the end of the chapter a 'Species profile' is given that summarises some of the salient features of the platypus and its biology.

General questions (Chapter 1)

Q. Is the platypus really shy?
A. The word 'secretive' is probably better than 'shy'. The platypus spends most of the daylight hours in its burrow and is active mainly at night. It is often difficult to see in the water as it blends in well and often forages under overhanging vegetation along the river bank. In captivity, platypuses are also normally secretive and may easily be stressed by human activity and handling. High mortality was a previous feature in zoos that housed the platypus. A few individuals have even died quite suddenly after being briefly handled, and veterinary examination shows no apparent reason for their deaths. However, zoos now have much better facilities, skilled husbandry practices and normally they only take juvenile animals and not established adults from the wild. In the past few years, these methods have resulted in much better survival rates of captive platypuses and greater success in captive breeding.

Q. What is the correct plural of the name platypus?
A. Platypus is apparently an anglicised Greek word meaning flat foot, presumably referring to the web of the foot, and therefore should have the English ending '-es'; i.e. 'platypuses'. Dictionaries also give 'platypus' and 'platypi' but the '-i' ending is Latin and really should not be used.

***Q. Why do some books have the scientific name* Ornithorhynchus paradoxus *instead of* Ornithorhynchus anatinus?**
A. When the first specimen was named in England it was given the name *Platypus anatinus* in 1799. Another biologist also received a specimen around the same time and in 1800 named it *Ornithorhynchus paradoxus*. Unfortunately the name 'Platypus' ('flat-foot') had already been applied to a genus of beetles and so had to be changed. *Ornithorhynchus* ('bird-snout') replaced it but, because *anatinus* ('duck-like') pre-dated the use of the *paradoxus* species name, *anatinus* was kept. Early settlers often called the animal the 'duck-mole' or 'water mole'. Neither of these common names has persisted but many people still use the term 'duck-billed platypus'.

Q. How long does a platypus live?
A. At least 21 years in the wild, although many marked animals have disappeared before reaching that age in wild populations, and are presumed to have died. The oldest animals recaptured in a long-term study in the upper Shoalhaven River in New South Wales have been females but some of the long-lived (around 20 years) captive animals have been males. However, in wild populations males are recaptured less often than females, so it has been more difficult to determine an accurate lifespan for them.

Q. Have platypuses ever been found in Western Australia?
A. No, there is no historical or current palaeontological evidence the species ever occurred naturally in Western Australia but there has been at least one unsuccessful introduction.

Q. Is the bill like that of a duck?
A. Not at all. Unlike a duck bill, the platypus bill is soft and pliable, like soft leather.

Breeding biology (Chapter 2)

Q. What do you call a baby platypus?
A. A few years ago a researcher on Kangaroo Island coined the name 'puggle' to apply to baby echidnas. The media also used this name for baby platypuses bred in captivity at Sydney's Taronga Zoo. Children at a kindergarten in Brisbane came up with the name 'plateena' using the

word 'pateenah', the word for egg in the language of the Nuenonne people of southern Tasmania. This was published in the *Australian Geographic Magazine* in 1999 but, like 'puggle', it does not seem to have caught on. In general, platypus keepers and researchers see no need to have a special name at all for a baby platypus. After all, there are no special names for baby wombats, koalas, bandicoots, numbats etc.

Q. Have platypuses been bred in captivity?
A. In the 1943/44 breeding season David Fleay, at the Sir Colin MacKenzie Sanctuary in Healesville near Melbourne, was involved in the first captive breeding of the platypus. Although several bred in large artificial pools at Warrawong Sanctuary in the Adelaide Hills between 1991 and 1996, the next successful breeding in a zoo was not until the 1998/99 breeding season. This captive breeding also occurred at Healesville Sanctuary, where another breeding took place two years later. Since then, Taronga Zoo in Sydney has also bred the platypus several times.

Q. How many young does the platypus have?
A. Between one and three, but no one knows how many young from an individual mother survive in the wild. Currently, two has been the maximum number of young produced in captivity.

Q. Do platypuses occur in male–female pairs?
A. There appears to be little evidence of any permanent male–female pairs.

Q. Do they occur in family groups?
A. There is no evidence that they do. Certainly in captivity the mother has little or nothing to do with the young a short time after they leave the nesting burrow.

Q. What is the sex ratio in platypus populations?
A. Although in certain studies the sex ratio of both adult and juvenile platypuses has not been significantly different from 1:1, in a long-term study in the upper Shoalhaven River in New South Wales, a significant bias towards females has been found in adults and juveniles. In 11 populations of varying sizes studied by the Australian Platypus Conservancy in Victoria, males were found to comprise from as little as 30% to as much as 70% of the population.

Q. How is sex determined in the platypus?
A. As in other mammals, sex is determined by the inheritance of X and Y chromosomes but rather than a single X and Y chromosome, the male platypus has 5 X chromosomes and 5 Y chromosomes. The female has a corresponding 5 pairs of X chromosomes.

Q. Is the male platypus involved in rearing the young?
A. Almost certainly not. In captivity the males are separated from the females once they have mated and so no data are available.

Q. What happens to the juvenile platypuses once they leave the nesting burrows?
A. Most independent juveniles appear to leave their natal area, although a certain number of females may stay and end up breeding. A much lower proportion of marked juvenile males than females are normally recaptured.

Spurs and venom glands (Chapter 3)

Q. Do both male and female platypuses have spurs?
A. No, only the males have spurs.

Q. Is there any way to handle a male platypus without getting spurred, say if you found one injured or caught one on a fishing line?
A. Yes, if you grasp the back half of the tail it cannot reach your hand with its spurs but you must be careful not to let your arm or other part of your body come close to the spurs, which are on the inside of its back legs. If it has a fish hook in its bill, wrap the platypus up in a Hessian bag or thick jacket so the hind legs are pinned against its body. Then you can remove the hook. Cutting the line and leaving the hook in place will probably hinder the animal from feeding and may result in death by starvation or infection.

Q. Can platypus venom kill a human?
A. No, but it can cause extreme pain, swelling and immobility of limbs and joints.

Q. Is it true that the venom can kill a dog?
A. Before platypuses were protected by law, there were reports of dogs dying from being spurred when they retrieved shot platypuses from the water during hunting. Nowadays a dog would seldom come into contact with a platypus.

Q. Are the spurs used by the male platypus to hold and subdue the female during mating?
A. It does not appear that the male uses the spurs purposely to hold the female during mating.

The sensory world of the platypus (Chapter 4)

Q. Is it true that the platypus closes its eyes, ears and nostrils when it is underwater?
A. Yes it does. The eyes and ears are housed in a furrow on the side of the head and these close when the animal dives. The nostrils are also closed by valves during diving.

Q. If they close their eyes, ears and nostrils, how do they find their food underwater?
A. Nobody really knows for sure, but there are sensitive touch receptors and electroreceptors in the bill that are thought to be involved in location of their prey.

Q. Do platypuses have colour vision?
A. They must have some colour vision, as there are cone cells in the retina, but which colours they see is not known.

Q. Do platypuses communicate with vocalisations?
A. Animals will make a low-pitched growling noise when they are being handled and a quiet snuffling noise can sometimes be heard from nest boxes housing platypuses in captivity. These vocalisations have not been reported in the wild, although noises were recorded from nestlings and featured in the Australian Broadcasting Corporation documentary: *Platypus. World's strangest animal.*

Q. Do platypuses sleep?
A. Yes they do. They even exhibit long periods of rapid eye movement (REM) sleep, which is associated with dreaming in humans. No one knows if platypuses dream!

Energetics, diving and foraging (Chapter 5)

Q. What do platypuses eat?
A. They eat a range of invertebrate species including insect larvae, small snails and freshwater mussels, shrimps and yabbies (freshwater crayfish), mainly from the bottom of the body of water they inhabit. These are called benthic macroinvertebrates.

Q. How long can a platypus stay underwater?
A. When a platypus exerts energy while foraging it only stays underwater for a maximum of just over two minutes. Normally a platypus will spend less than one minute foraging along the bottom followed by around 10–15 seconds on the surface between normal dives. In captivity (and possibly occasionally in the wild) an individual can stay down for about 10 minutes if it wedges itself under a log or rock. In this situation it is inactive and does not use up as much oxygen.

Q. How deep can a platypus dive?
A. Nobody really knows, but one was caught in a net set 30 metres deep in Lake Yarrunga near Kangaroo Valley in New South Wales. They often seem to avoid foraging in very shallow water (less than one metre in depth) but have been reported diving to nearly 9 metres to obtain their food. Once the water gets too deep, say over 10 metres, the energy used getting that far down to obtain small food items would make it unlikely that they would be able forage regularly at those depths.

Q. Its prey items are very small. How often does a platypus have to dive during a feeding period?
A. On average, platypuses dive about 75 times per hour. Radio-tracked animals were recorded in Lake Lea in Tasmania diving up to 1600 times during a foraging period.

Q. How long each day does a platypus spend feeding?
A. A very long time indeed, especially during winter when its energy demands are higher in order to maintain its body temperature. On average a platypus spends about 12 hours a day feeding in the water but radio-tracked animals have been followed foraging continuously for up to 30 hours.

Q. Is it true that platypuses are only found in the river around dawn and dusk?
A. Being active at dawn and dusk is called crepuscular. However, platypuses aren't really crepuscular. Most forage throughout the night (nocturnal). People do see them most often at dusk and dawn but that is just because they start to come out of their burrows an hour or so before it gets dark and return over a similar period of time after first light.

Q. How fast do platypuses swim?
A. Platypuses have been recorded swimming through riffles (rapids) flowing at around 1 metre per second (3.6 km per hour) but during normal foraging they swim at around 0.4 metres per second (1.45 km per hour).

Q. What is the normal body temperature of the platypus?
A. At 32°C, the normal body temperature of the platypus is a bit lower than that of a human and most other mammals but it is maintained constantly close to that level. Echidnas have a similar body temperature to platypuses but are also known to reduce that temperature significantly during hibernation.

Q. Do platypuses hibernate?
A. No platypuses exhibited hibernation during three radio-monitoring studies in sub-alpine or tableland areas in winter. A certain number have been inactive for nearly seven days in captivity and in the wild but, as their temperatures were not being measured at the time, it is impossible to tell whether or not this inactivity was hibernation.

Ecology (Chapter 6)

Q. How many platypuses are there in a stream, an area, state or even the whole of Australia?

A. Because of its secretive habits, mobility and penchant for avoiding capture in mark–recapture studies, actual platypus numbers are very difficult to assess. There are almost certainly thousands of them in each of the states in which they are normally found (Queensland, New South Wales, Victoria and Tasmania).

Q. What predators do platypuses have?
A. As platypuses are normally either foraging in the water or tucked up in their burrows, it is difficult to determine what might kill and eat them. Foxes and possibly dogs or dingoes take platypuses moving overland or in shallow water (Chapter 8), occasionally also digging them out of their burrows. Early naturalists suggested crocodiles, goannas, carpet pythons, eagles and large native fish, like Murray cod, were natural predators. There are reports of dead platypuses being carried by individual wedge-tailed eagles, one was apparently vomited up by an eastern cod (*Maccullochella ikei*) being handled by fisheries scientists, and at least one was reportedly found in the stomach of a crocodile. At Warrawong Sanctuary in South Australia, captive native water rats (*Hydromys chrysogaster*) inflicted bites to the feet and tails of female platypuses when they were housed in the same ponds. Whether this occurs in the wild is not known but suggestions have been made that water rats may take platypus nestlings from nesting burrows. Most reports of predation are anecdotal and there is even a report of a platypus being found in the stomach of an Australian fur seal (*Arctocephalus pusillus*) in the lower Shoalhaven River in New South Wales.

Q. What sort of habitat does the platypus prefer?
A. Platypuses are found over a wide range of habitats, ranging from quite pristine streams to highly degraded ones. However, a 'platypus heaven' is a river or stream with earth banks consolidated by the roots of native vegetation, the foliage of which supplies shade and cover for platypuses foraging and entering burrows. The stream itself contains logs, twigs, roots, water plants and has a gravel and cobble bottom. All of these are important to invertebrate animals eaten by the platypus and the presence of both pools and riffles (rapids) provide extensive and varied foraging areas for these invertebrates.

Q. What is the home range of the platypus?
A. Some individual platypuses will forage up to 10 km in a 24-hour period, while others seem to be more restricted in their movements. Reported home ranges for individuals vary between 200 metres and 7 km of stream. Dispersing juvenile animals move over much greater distances.

Q. What happens to platypuses during droughts?
A. Platypuses must feed in water and so cannot survive in completely dry streams. Some do however survive during dry conditions by moving to refuge pools which retain water in which food is available. There is almost certainly considerable mortality as a result of drought, and prolonged drought conditions predicted over parts of the species' current distribution is of concern and could result in local extinctions.

Q. What do they do during floods?
A. They certainly have been found before and after floods in many places, including those affected by wet seasons and even cyclones. Under these conditions, platypuses tend to feed in the less rapidly flowing water along river banks or in flooded pools adjacent to the streams, and may seek refuge in other shelters if their burrows are underwater. Limited evidence suggests that significant mortality of nest-dependent young may occur if flooding happens during the breeding season.

Ancestry and evolution (Chapter 7)

Q. Is it true that the platypus was swimming around the feet of dinosaurs in Australia?
A. No. The modern platypus has been in Australia for only about five million years and the dinosaurs had died out in Australia by around 65 million years ago. However, there are three other monotremes in the Australian fossil record which may have been platypus-like in appearance from as long ago as 110 million years.

Q. Have any fossil platypuses been found on other continents?
A. Yes. Three teeth belonging to a fossil platypus species (*Monotrematum sudamericanum*) have been found in Patagonia, Argentina.

Q. Did the ancestors of the platypus originate in Australia?
A. Nobody really knows, but the greater diversity of fossil species found in Australia than in South America suggests that their ancestors may have had an Australian origin.

Conservation: platypuses and people (Chapter 8)

Q. Is the platypus protected?
A. Yes, in the Australian Capital Territory and in every state in which the platypus occurs, it is protected by legislation. This means that platypuses cannot be captured or killed, except for scientific research, and can be kept only in licensed zoos or sanctuaries. Special permits and ethics approval are necessary before any platypuses can be captured or killed for research or display. Only the Federal Government wildlife authorities can approve the export of live platypuses or scientific specimens (e.g. blood or tissue) taken from platypuses.

Q. What is the conservation status of the platypus? Is it threatened or endangered?
A. No. In 2006 the International Union for the Conservation of Nature and Natural Resources (IUCN) classified the platypus as a 'species of least concern'. Apart from its probable extinction in the Adelaide Hills and Mount Lofty Ranges in South Australia, the platypus still occupies the majority of its original distribution. However, there is evidence that its distribution has been fragmented in certain river systems resulting in local population declines and even extinctions.

Q. Are there any threats to platypus conservation?
A. Inappropriate land and water management, both today and in the past, have affected its habitat and represent the major long-term threats to the species because of its dependence on established banks in which to burrow and on the water for its food. Illegal fishing using traps, nets and set lines are an immediate real threat to local populations, especially in small streams where the actual numbers may normally be quite low. Drowning one or more breeding animals could nudge the population towards local extinction.

Q. Is it true that platypuses are drowned in fishing nets and traps, such as yabby (freshwater crayfish) traps?
A. Yes it is. Because they can hold their breath for only a short time when they are trapped and struggling, they drown in less than five minutes. Yabby traps are perhaps the worst, as platypuses eat yabbies and so are probably attracted into the traps. Yabby traps, including the commonly available 'opera house traps' are now banned in most states in streams where platypuses occur, except for Queensland. Most other nets and traps are also illegal in freshwaters in all states.

Q. Is there anything I can do about these threats?
A. Yes there is. It is essential to let the local wildlife and / or fisheries authorities know if you do find nets, traps or set lines. As an individual, the much bigger issues of poor land and water management are more difficult to tackle. However, it is important to be aware of land and water issues in your area and to lobby the local, state and federal governments about projects and activities that affect the platypus. You can also get involved in Landcare, Bushcare and Rivercare rehabilitation projects. There is plenty of information available on ways to preserve or rehabilitate streams, lakes and even farm dams to make them more suitable for use by the platypus and other wildlife species. In most states, one or more of the bodies responsible for land and water conservation have local Landcare cocoordinators or equivalent staff. One of these people is the best place to start to look for advice.

Q. What do I do if I find an injured platypus?
A. Platypuses are normally either in their burrows in the bank during the day or in the water feeding, mainly at night. However, there are those that do venture away from the water. This happens most often during the autumn, when juveniles may be moving out of their natal area to find their own home ranges. Platypuses may also get displaced during droughts or floods, when their normal home ranges have been disturbed. If an animal appears to be alert and active, although in the wrong place, it is better to take it to the nearest waterway and let it go. They are very difficult to keep in captivity and so should not be kept unless they appear to be injured or unwell. In this case an injured or unwell animal must be taken to a veterinarian or reported to one of the animal rescue services that will then pass it on to a zoo or sanctuary experienced in handling platypuses. Because

of their special amphibious lifestyle, the platypus is very difficult to look after at home, even for an experienced wildlife carer.

If you need to keep the animal until you can get it to the veterinarian, zoo or sanctuary, there are a few things to do and not to do:

- **ALWAYS** pick it up by the end half of the tail. **DON'T** put your hand under it to support it or get any other part of your body near the spurs, which are on the inside of its back legs. Assume it is a male until you find out otherwise.
- **DO** put it into a sturdy container (but not airtight), or better, into a strong cloth bag (e.g. pillow case or garment with the sleeves knotted), tie the opening tight but leave the bag as loose as possible so the animal can move around.
- **DO** put the bag in a dry, quiet and dark place away from any disturbance.
- **DON'T** be tempted to show other people what you have found. Platypuses have been known to die quickly from too much disturbance and stress.
- **DON'T** put the animal into water.
- **DON'T** try to give it water to drink. The animal may be dehydrated but must be rehydrated by a veterinarian or qualified person.
- **DON'T** attempt to feed it anything. A platypus normally has at least some stored tail fat that will be enough of a food source for it until you get to the veterinarian, zoo or sanctuary.
- **DON'T** keep it too warm. Platypuses have a lower body temperature than humans and will die from heat stress above 30°C. An ideal temperature is **20–25°C**.
- **DO** seek veterinary assistance **absolutely as soon as possible**.

Summary: Species profile

Genus: *Ornithorhynchus*

Species: *anatinus*

Size: Males 400–630 mm, 800–3000 g; Females 370–550 mm, 600–1700 g

Distribution: Rivers, creeks and lakes of eastern Australia from Cooktown to Tasmania

Status: Common or 'of least concern' (IUCN) but locally reduced in numbers or even extinct from certain streams

Body temperature: 32°C; homeothermic

Food: Mainly benthic invertebrates, especially insect larvae

Mating: July–October over most of mainland Australia but as late as February in Tasmania

Gestation: Approximately 21 days

Incubation: Unknown. Possibly 10 days

Lactation: Three to four months; possibly over four months in captivity

Emergence of young: Late January–April over most of mainland Australia but as late as May in Tasmania

Longevity: 20 years in captivity; up to 21 years in wild

Mortality: Poorly known:

- Predation by foxes and occasionally birds of prey, crocodiles and large freshwater fish species (e.g. eastern cod). Possibly large eels, goannas and water rats.
- Starvation and / or heat stress in dispersing juveniles and / or adults during drought.
- Human activities, especially drowning in illegal nets and traps and becoming entangled in discarded fishing line, plastic and other rubbish.
- *Mucor* fungal disease in Tasmania.

SELECTED REFERENCES

Chapter 1. Introduction

1. Abbott I (2008). Historical perspectives of the ecology of some conspicuous vertebrate species in south-western Australia. *Nature Conservation Western Australia* **6** (in press).
2. Augee M, Gooden B and Musser A (2006). *Echidna. Extraordinary Egg-laying Mammal*. CSIRO Publishing, Melbourne.
3. Burnett S (2006). Platypus in the collective consciousness. The Queensland surveys. *Wildlife Australia Magazine* **43**, 33–36.
4. Flannery TF and Groves CP (1998). A revision of the genus *Zaglossus* (Monotremata, Tachyglossidae), with descriptions of a new species and subspecies. *Mammalia* **62**, 367–396.
5. Fleay D (1980). *Paradoxical Platypus: Hobnobbing with Duckbills*. Jacaranda Press, Milton, Queensland.
6. Grant TR (1992). Historical and current distribution of the platypus, *Ornithorhynchus anatinus*, in Australia. In *Platypus and Echidnas*. (Ed ML Augee) pp. 232–254. Royal Zoological Society of NSW, Sydney.
7. Grant TR (2004). Depth and substrate selection by platypuses, *Ornithorhynchus anatinus*, in the lower Hastings River, New South Wales. *Proceedings of the Linnean Society of NSW* **125**, 235–241.
8. Grant TR and Bishop K (1998). Instream flow requirements for the platypus (*Ornithorhynchus anatinus*) – a review. *Australian Mammalogy* **20**, 267–280.
9. Grant TR and Temple-Smith PD (1998). Field biology of the platypus (*Ornithorhynchus anatinus*) – historical and current perspectives. *Transactions of the Royal Society of London Series B* **353**, 1081–1091.
10. Grant TR and Temple-Smith PD (2003). Conservation of the platypus, *Ornithorhynchus anatinus*: threats and challenges. *Aquatic Ecosystem Health and Management* **6**, 5–18.
11. Griffiths M (1978). *The Biology of the Monotremes*. Academic Press, New York.
12. Lester KS and Archer M (1986). A description of the molar enamel of a middle Miocene monotreme (*Obdurodon*, Ornithorhynchidae). *Anatomy and Embryology* **174**,145–151.
13. Lester KS, Boyde A, Gilkeson C and Archer M (1987). Marsupial and monotreme enamel structure. *Scanning Microscopy* **1**, 401–420.
14. Otley HM, Munks SA and Hindel MA (2000). Activity patterns, movements and burrows of platypuses (*Ornithorhynchus anatinus*) in a sub-alpine Tasmanian lake. *Australian Journal of Zoology* **48**, 701–713.
15. Pridmore PA (1985). Terrestrial locomotion in monotremes (Mammalia: Monotremata). *Journal of Zoology* (London) **205**, 53–74.
16. Robinson AC, Casperson KD and Hutchinson MN (2002). A list of the vertebrates of South Australia. Biological Survey and Research, Heritage and Biodiversity Division, Department for Environment and Heritage, Adelaide.
17. Serena M (1994). Use of time and space by the platypus (*Ornithorhynchus anatinus*) along a Victorian stream. *Journal of Zoology* (London) **232**, 117–131.

18. Serena M, Thomas JL, Williams GA and Officer RCE (1998). Use of stream and river habitats by the platypus, *Ornithorhynchus anatinus*, in an urban fringe habitat. *Australian Journal of Zoology* **46**, 267–282.
19. Serena M, Worley M, Swinnerton M and Williams GA (2001). Effect of food availability and habitat on the distribution of platypus (*Ornithorhynchus anatinus*) foraging activity. *Australian Journal of Zoology* **49**, 263–277.
20. Stone GC (1983). Distribution of the platypus (*Ornithorhynchus anatinus*) in Queensland. Queensland National Parks and Wildlife Service, Brisbane.
21. Williams G and Serena M (1999). Living with platypus. A practical guide to the conservation of a very special Australian. Australian Platypus Conservancy, Whittlesea, Victoria.

Chapter 2. Breeding biology

[Also see references 5, 10, 14].
22. Anonymous (2006). Juveniles boost reintroduction. *Ripples* Issue **33**, 1. Newsletter of the Australian Platypus Conservancy, Whittlesea, Victoria.
23. Ashley T (2005). Chromosome chains and platypus sex: kinky connections. *Bioessays* **27**, 681–684.
24. Bennett G (1835). Notes on the natural history and habits of the *Ornithorhynchus paradoxus*, Blum. *Transactions of the Zoological Society of London* **1**, 229–258.
25. Burrell H (1927). *The Platypus*. Angus and Robertson, Sydney.
26. Caldwell WH (1887). The embryology of Monotremata and Marsupialia, Part I. *Philosophical Transactions of the Royal Society of London Series B* **178**, 463–486.
27. Carrick FN and Hughes RL (1978). Reproduction in male monotremes. *Australian Zoologist* **20**, 211–231.
28. Carrick FN and Hughes RL (1982). Aspects of the structure and development of monotreme spermatozoa and their relevance to the evolution of mammalian sperm morphology. *Cell and Tissue Research* **222**, 127–141.
29. Chisholm AH (1959). Scientists wrangle over platypus puzzle. *Sydney Morning Herald*, 4 July 1959.
30. Connolly JH and Obendorf DL (1998). Distribution, captures and physical characteristics of the platypus (*Ornithorhynchus anatinus*) in Tasmania. *Australian Mammalogy* **20**, 231–237.
31. Crawley TW and Company (1884). The duck-billed platypus: *Ornithorhynchus*. Letter to the *Sydney Morning Herald*, 20 September 1884.
32. De-La-Warr M and Serena M (1999). Observations of platypus *Ornithorhynchus anatinus* mating behaviour. *Victorian Naturalist* **116**, 172–174.
33. Fleay D (1944). *We Breed the Platypus*. Robertson and Mullens, Melbourne.
34. Gibson RA, Neumann M, Grant TR and Griffiths M (1988). Fatty acids of the milk and food of the platypus (*Ornithorhynchus anatinus*). *Lipids* **23**, 377–379.
35. Gilfedder L, Whinam J and Harris S (1992). An observation of apparent platypus nesting behaviour. *Tasmanian Naturalist* **109**, 4.
36. Grant TR and Temple-Smith PD (1983). Size, seasonal weight change and growth in platypuses, *Ornithorhynchus anatinus* (Monotremata: Ornithorhynchidae), in rivers and lakes of New South Wales. *Australian Mammalogy* **6**, 51–60.

37. Grant TR and Temple-Smith PD (1998). Growth of nestling and juvenile platypuses (*Ornithorhynchus anatinus*). *Australian Mammalogy* **20**, 221–230.
38. Grant TR (2004). Captures, capture mortality, age and sex ratios of platypuses, *Ornithorhynchus anatinus*, during studies over 30 years in the upper Shoalhaven River in New South Wales. *Proceedings of the Linnean Society of NSW* **125**, 217–226.
39. Grant TR, Griffiths M and Temple-Smith PD (2004). Breeding in a free-ranging population of platypuses, *Ornithorhynchus anatinus*, in the upper Shoalhaven River in New South Wales – a 27 year study. *Proceedings of the Linnean Society of NSW* **125**, 227–234.
40. Hamilton-Smith E (1968). Platypus in caves. *Victorian Naturalist* **85**, 292.
41 Griffiths M, Elliott MA, Leckie RMC and Schoefl GI (1973). Observations of the anatomy and ultrastructure of mammary glands and on fatty acids of the triglycerides in platypus and echidna milk fats. *Journal of Zoology* (London) **169**, 255–279.
42. Griffiths M, Green B, Leckie RMC, Messer M and Newgrain KW (1984). Constituents of platypus and echidna milk, with particular reference to the fatty acid complement of the triglycerides. *Australian Journal of Biological Science* **37**, 323–329.
43. Gruetzner F, Rens W, Tsend-Ayush E, El-Mogharbel N, O'Brien PCM, Jones RC, Ferguson-Smith MA and Graves JAM (2004). In the platypus a meiotic chain of ten sex chromosomes shares genes with the bird Z and mammal X chromosomes. *Nature* **432**, 913–917.
44. Handasyde KA, McDonald IR and Evans BK (1992). Seasonal changes in plasma concentrations of progesterone in free-ranging platypus (*Ornithorhynchus anatinus*). In *Platypus and Echidnas*. (Ed ML Augee) pp. 75–79. Royal Zoological Society of NSW, Sydney.
45. Handasyde KA, McDonald IR and Evans BK (2003). Plasma corticoid concentrations in free-ranging platypuses (*Ornithorhynchus anatinus*): response to capture and patterns in relation to reproduction. *Comparative Biochemistry and Physiology* **136A**, 895–902.
46. Hawkins M and Fanning D (1992). Courtship and mating of captive platypuses at Taronga Zoo. In *Platypus and Echidnas*. (Ed ML Augee) pp. 106–114. Royal Zoological Society of NSW, Sydney.
47. Holland N and Jackson SM (2002). Reproductive behaviour and food consumption associated with captive breeding of platypus (*Ornithorhynchus anatinus*). *Journal of Zoology* (London) **256**, 279–288.
48. Hughes RL and Carrick FN (1978). Reproduction in female monotremes. *Australian Zoologist* **20**, 233–253.
49. Jabukowski JM, New NP, Stone GM and Jones RC (1998). Reproductive seasonality in female platypuses, *Ornithorhynchus anatinus*, in the Upper Barnard River, New South Wales. *Australian Mammalogy* **20**, 207–213.
50. Jackson SM, Grant TR, Temple-Smith PD and Bennett G (2005). Monsieur Jules Verreaux and the platypus (*Ornithorhynchus anatinus*): comments on the observations of a naturaliste voyageur. *Australian Mammalogy* **27**, 147–160.
51. Joseph R and Griffiths M (1992). Whey proteins in early and late milks of monotremes (Monotremata: Tachyglossidae, Ornithorhynchidae) and of the tammar wallaby (*Macropus eugenii*, Marsupialia: Macropodidae). *Australian Mammalogy* **15**, 125–127.

52. Krueger B, Hunter S and Serena M (1992). Husbandry, diet and behaviour of platypus *Ornithorhynchus anatinus*. *International Zoo Yearbook* **31**, 64–71.
53. Lichon M (1999). Platypus in Tasmanian caves, and particularly in Croesus Cave. *Illuminations* (Journal of Mole Creek Caving Club) **4**, 21–27.
54. Maule L (1832). Habits and economy of the Ornithorhynchus. *Proceedings of the Committee of Science and Correspondence of the Zoological Society of London*. **Part II**, 145–148.
55. Messer M, Gadiel PA, Ralston GB and Griffiths M (1983). Carbohydrates of the milk of the platypus. *Australian Journal of Biological Science* **36**, 129–137.
56. Moyal A (2001). *Platypus: The Extraordinary Story of How a Curious Creature Baffled the World*. Allen and Unwin, Crows Nest.
57. Munks SA, Otley HM, Bethge P and Jackson J (2000). Reproduction, diet and daily energy expenditure of the platypus in a sub-alpine Tasmanian lake. *Australian Mammalogy* **21**(2), 260–261 Abstract.
58. Munks S, Eberhard R and Duhig N (2004). Nests of the platypus *Ornithorhynchus anatinus* in a Tasmanian cave. *Tasmanian Naturalist* **126**, 55–58.
59. New NP, Jabukowski JM, Stone GM and Jones RC (1998). Seasonal pattern of androgen secretion in the male platypus, *Ornithorhynchus anatinus*, in the Upper Barnard River, New South Wales. *Australian Mammalogy* **20**, 215–220.
60. Rismiller PD and McKelvey MW (2000). Frequency of breeding and recruitment in the short-beaked echidna, *Tachyglossus aculeatus*. *Journal of Mammalogy* **81**, 1–17.
61. Strahan R and Thomas D (1975). Courtship of the platypus, *Ornithorhynchus anatinus*. *Australian Zoologist* **18**, 165–178.
62. Temple-Smith PD (1973). Seasonal breeding biology of the platypus, *Ornithorhynchus anatinus* (Shaw 1799) with special reference to the male. PhD Thesis. Australian National University, Canberra.
63. Temple-Smith PD and Grant TR (2002). Uncertain breeding: A short history of reproduction in monotremes. *Reproduction, Fertility and Development* **13**, 487–497.

Chapter 3. Spurs and venom glands

[Also see references 11, 25, 32, 46, 47, 50, 61, 62].
64. Akiyama S (1999). Molecular ecology of the platypus (*Ornithorhynchus anatinus*). PhD Thesis. La Trobe University, Australia.
65. de Plater GM, Milburn PJ and Martin RL (2001). Venom of the platypus, *Ornithorhynchus anatinus*, induces a calcium-dependent current in cultured dorsal root ganglion cells. *Journal of Neurophysiology* **85**, 1340–1345.
66. Fenner PJ, Williamson JA and Meyers D (1992). Platypus envenomation – a painful experience. *The Medical Journal of Australia* **157**, 829–832.
67. Gardner JL and Serena M (1995). Spatial organisation and movement patterns of adult male platypus, *Ornithorhynchus anatinus* (Monotremata: Ornithorhynchidae). *Australian Journal of Zoology* **43**, 91–103.
68. Gemmell NJ, Grant TR, Western PS, Wamsley J, Watson JM, Murray ND and Graves JAM (1995). Determining platypus relationships. *Australian Journal of Zoology* **43**, 283–291.

69. Gust N and Handasyde K (1995). Seasonal variation in the ranging behaviour of the platypus (*Ornithorhynchus anatinus*) on the Goulburn River, Victoria. *Australian Journal of Zoology* **43**, 193–208.
70. Ji Q, Luo Z-X, Yuan C-X and Tabrum AR (2006). A swimming mammaliform from the middle Jurassic and ecomorphological diversification of early mammals. *Science* **311**, 1123–1127.
71. Jamison J (1818). Extract from the Minute Book of the Society. Letter to the Secretary written 10 September 1816. *Transactions of the Linnean Society of London* **12**, 584–585.
72. Kielan-Jaworowska Z and Hurum JH (2006). Limb posture in early mammals: sprawling or parasagittal. *Acta Palaeontoligica Polonica* **51**, 393–406.
73. Sutherland SK and Timballs J (2001). *Australian Animal Toxins: The Creatures, Their Toxins and Care of the Poisoned Patient*. 2nd edition. Oxford University Press, Melbourne.
74. Tonkin MA and Negrine J (1994). Wild platypus attack in the Antipodes – a case report. *Journal of Hand Surgery [British and European]* **19B**, 162–164.
75. Torres AH and Kuchel PW (2000). The platypus and its venom. *The Biochemist* February, 33–36.
76. Torres AM and Kuchel PW (2004) The ß-defensin-fold family of polypeptides. *Toxicon* **44**, 581–588.
77. Torres AM, Tsampazi M, Tsampazi C, Kennett EC, Belov K, Geraghty DP, Bansal BS, Alewood PF and Kuchel PW (2006). Mammalian L-to-D-amino-acid-residue isomerase from platypus. *FEBS Letters* **580**, 1587–1591.

Chapter 4. The sensory world of the platypus

[Also see references 2, 11, 25, 61, 63].
78. Andres KH and von During M (1988). Comparative anatomy of vertebrate electroreceptors. In *Progress in Brain Research*. (Eds W Hamann and A Iggo) Volume 74. Chapter 14. pp. 103–113. Elsevier, Amsterdam.
79. Bohringer RC (1992). The platypus bill receptors and their central connections. In *Platypus and Echidnas*. (Ed ML Augee) pp. 194–203. Royal Zoological Society of NSW, Sydney.
80. Bohringer RC and Rowe MJ (1977). The organisation of the sensory and motor areas of the cerebral cortex in the platypus (*Ornithorhynchus anatinus*). *Journal of Comparative Neurology* **174**, 1–14.
81. Gates GR, Saunders JC and Boek GR (1974). Peripheral auditory function in the platypus, *Ornithorhynchus anatinus*. *Journal of the Acoustical Society of America* **56**, 152–156.
82. Gunn M (1884). On the eye of *Ornithorhynchus paradoxus*. *Journal of Anatomy and Physiology. Normal and Pathological*. **18**, 400–405.
83. Ladham A and Pickles JO (1996). Morphology of the organ of Corti and macula lagena. *Journal of Comparative Neurology* **366**, 335–347.
84. Macrini TE, Rowe T and Archer M (2006). Description of a cranial endocast from a fossil platypus, *Obdurodon dicksoni* (Monotremata, Ornithorhynchidae), and the relevance of endocranial characters to monotreme monophyly. *Journal of Morphology* **267**, 1000–1015.

85. Manger PR and Pettigrew JD (1995). Electroreception and the feeding behaviour of platypus (*Ornithorhynchus anatinus*: Monotremata: Mammalia). *Philosophical Transactions of the Royal Society of London Series B* **347**, 359–381.
86. Pettigrew JD, Manger PR and Fine SLB (1998). The sensory world of the platypus. *Philosophical Transactions of the Royal Society of London Series B* **353**, 1199–1210.
87. Poulton EB (1885). On the tactile terminal organs and other structures in the bill of *Ornithorhynchus*. *Journal of Physiology* **5**, 15–16.
88. Proske U and Gregory E (2003). Electrolocation in the platypus – some speculations. *Comparative Biochemistry and Physiology Part B* **136**, 821–825.
89. Proske U and Gregory JE (2004). The role of push rods in the platypus and echidna – some speculations. *Proceedings of the Linnean Society of NSW* **125**, 319–326.
90. Rich TH, Hopson JA, Musser AM, Flannery TF and Vickers-Rich P (2005). Independent origins of middle ear bones in Monotremes and Therians. *Science* **307**, 910–914.
91. Rismiller PD (1992). Field observations on the Kangaroo Island echidnas (*Tachyglossus aculeatus multiaculeatus*) during the breeding season. In *Platypus and Echidnas*. (Ed ML Augee) pp. 101–105. Royal Zoological Society of NSW, Sydney.
92. Rowe MJ, Mahns DA and Sahai V (2004). Monotreme tactile mechanisms: from sensory nerves to cerebral cortex. *Proceedings of the Linnean Society of New South Wales* **125**, 301–317.
93. Scheich H, Langer G, Tideman C, Coles RB and Guppy A (1986). Electroreception and electrolocation in platypus. *Nature* **319**, 401–402.
94. Siegel JM, Manger PR, Nienhuis R, Fahringer HM and Pettigrew JD (1998). Monotremes and the evolution of rapid eye movement sleep. *Philosophical Transactions of the Royal Society of London Series B* **353**, 1147–1157.
95. Taylor NG, Manger PR, Pettigrew JD and Hall LS (1992). Electromyogenic potentials of a variety of platypus prey items: an amplitude and frequency analysis. In *Platypus and Echidnas*. (Ed ML Augee) pp. 216–224. Royal Zoological Society of NSW, Sydney.
96. Wilson JT and Martin CJ (1893). On the peculiar rod-like tactile organs in the integument and mucous membrane of the muzzle of *Ornithorhynchus*. *Linnean Society of New South Wales*. Macleay Memorial Volume, pp. 190–200.
97. Wilson JT and Martin CJ (1894). Further observations on the anatomy of the integumentary structures in the muzzle of *Ornithorhynchus*. *Linnean Society of New South Wales Series* **2**, 660–681.
98. Young HM and Vaney DL (1990). The retinae of Prototherian mammals possess neuronal types that are characteristic of non-mammalian retinae. *Visual Neuroscience* **5**, 61–66.

Chapter 5. Energetics, diving and foraging

[Also see references 5, 7, 8, 14, 15, 18, 19, 25, 47, 52, 57, 67, 69, 71, 85, 127].
99. Allport M (1878). The platypus. The following notes on the platypus (*Ornithorhynchus anatinus*). *Papers and Proceedings of the Royal Society of Tasmania*, 30–31.
100. Anonymous (1948). Platypus on dry fly. *Wild Life* June, 277.

101. Anonymous (1996). Mullum Mullum Creek update. *Ripples* Issue **6**, 3. September. Newsletter of the Australian Platypus Conservancy, Whittlesea, Victoria.
102. Bennett G (1835). Notes on the natural history and habits of the *Ornithorhynchus paradoxus*, Blum. *Transactions of the Zoological Society of London* **1**, 229–258.
103. Bethge P (2002). Energetics and foraging behaviour of the platypus, *Ornithorhynchus anatinus*. PhD Thesis. University of Tasmania, Australia.
104. Bethge P, Munks S and Nicol S (2001). Energetics of foraging and locomotion in the platypus, *Ornithorhynchus anatinus*. *Journal of Comparative Physiology – B, Biochemical, Systematic and Environmental Physiology* **171**, 497–506.
105. Bethge P, Munks S, Otley H and Nicol S (2003). Diving behaviour, dive cycles and aerobic dive limit in the platypus, *Ornithorhynchus anatinus*. *Comparative Biochemistry and Physiology* **136A**, 799–809.
106. Bethge P, Munks S, Otley H and Nicol S (2004). Platypus burrow temperatures at a subalpine Tasmanian lake. *Proceedings of the Linnean Society of NSW* **125**, 272–276.
107. Dawson TJ and Fanning FD (1981). Thermal and energetic problems of semi-aquatic mammals, a study of the Australian water rat, including comparisons with the platypus. *Physiological Zoology* **54**, 285–296.
108. Eadie R (1935). Hibernation in the platypus. *Victorian Naturalist* **52**, 71–72.
109. Evans BK, Jones DR, Baldwin J and Gabbott GRT (1994). Diving ability in the platypus. *Australian Journal of Zoology* **42**, 17–27.
110. Faragher RA, Grant TR and Carrick FN (1979). Food of the platypus, *Ornithorhynchus anatinus*, with notes on the food of the brown trout, *Salmo trutta*, in the Shoalhaven River, New South Wales. *Australian Journal of Ecology* **4**, 171–179.
111. Fish FE, Baudinette RV, Frappell PD and Sarre MP (1997). Energetics of swimming by the platypus *Ornithorhynchus anatinus*: effort associated with rowing. *Journal of Experimental Biology* **200**, 2647–2652.
112. Fish FE, Frappell PB, Baudinette RV and MacFarlane PM (2001). Energetics of terrestrial locomotion of the platypus, *Ornithorhynchus anatinus*. *Journal of Experimental Biology* **204**, 797–803.
113. Frappell PB (2003). Ventilation and metabolic rate in the platypus: insights into the evolution of the mammalian breathing pattern. *Comparative Biochemistry and Physiology* **136A**, 943–953.
114. Grant TR (1982). Food of the platypus, *Ornithorhynchus anatinus* (Monotremata: Ornithorhynchidae) from various water bodies in New South Wales. *Australian Mammalogy* **5**, 135–136.
115. Grant TR (1983). Body temperature of free-ranging platypuses, *Ornithorhynchus anatinus* (Monotremata), with observations on their use of burrows. *Australian Journal of Zoology* **31**, 117–122.
116. Grant TR and Dawson TJ (1978a). Temperature regulation in the platypus, *Ornithorhynchus anatinus*, maintenance of body temperature in air and water. *Physiological Zoology* **51**, 1–6.
117. Grant TR and Dawson TJ (1978b). Temperature regulation in the platypus, *Ornithorhynchus anatinus*, production and loss of metabolic heat in air and water. *Physiological Zoology* **51**, 315–332.
118. Grigg G, Beard L, Grant T and Augee M (1992). Body temperature and diurnal activity patterns in the platypus (*Ornithorhynchus anatinus*) during winter. *Australian Journal of Zoology* **40**, 135–142.

119. Harrop CJF and Hume ID (1980). Digestive tract and digestive function in monotremes and nonmacropod marsupials. In *Comparative Physiology: Primitive Mammals*. (Eds K Schmidt-Nielsen, L Bolis and CR Taylor) pp. 63–77. Cambridge University Press, Cambridge, UK.
120. Hulbert AJ and Grant TR (1983). A seasonal study of body condition and water turnover in a free-ranging population of platypuses, *Ornithorhynchus anatinus* (Monotremata). *Australian Journal of Zoology* **31**, 109–116.
121. Kruuk H (1993). The diving behaviour of the platypus (*Ornithorhynchus anatinus*) in waters with different trophic status. *Journal of Applied Ecology* **30**, 592–598.
122. McLachlan-Troup TA (2007) The ecology and functional importance of the platypus (*Ornithorhynchus anatinus*) in Australian freshwater habitats. PhD Thesis. University of Sydney, Australia.
123. Martin CJ (1902). Thermal regulation and respiratory exchange in monotremes and marsupials. A study in the development of homeothermism. *Philosophical Transactions of the Royal Society of London Series B* **195**, 1–37.
124. Munks SA, Otley HM, Bethge P and Jackson J (2000). Reproduction, diet and daily energy expenditure of the platypus in a sub-alpine Tasmanian lake. *Australian Mammalogy* **21**(2), 260–261 Abstract.
125. Nicholls AG (1958). The population of a trout stream and the survival of released fish. Appendix II. The natural predators of trout in Tasmania. *Australian Journal of Marine and Freshwater Research* **9**, 320–350.
126. Olsen PD (1995). Water rat *Hydromys chrysogaster*. In *The Mammals of Australia*. (Ed R Strahan) pp. 628–629. Australian Museum / Reed Books, Sydney.
127. Robinson KW (1954). Heat tolerance of Australian monotremes and marsupials. *Australian Journal of Biological Science* **7**, 248–360.
128. Semon R (1899). *In the Australian Bush*. Macmillan, London.
129. Sutherland A (1897). Temperatures of reptiles, monotremes and marsupials. *Proceedings of the Royal Society of Victoria* **9**, 57–67.
130. Williams WR (1945). Letter to the editor. *Wild Life* October, 312.

Chapter 6. Ecology

[Also see references 5, 7, 8, 9, 14, 17, 18, 19, 24, 25, 26, 38, 39, 67, 68, 69].
131. Anonymous (1999). Keeping tabs on a marathon swimmer. *Ripples* Issue **13**, 1. Newsletter of the Australian Platypus Conservancy, Whittlesea, Victoria.
132. Anonymous (2001). Platypus on the move. *Ripples* Issue **18**, 1. Newsletter of the Australian Platypus Conservancy, Whittlesea, Victoria.
133. Anonymous (2004). Platypus on the move. *Ripples* Issue **27**, 3. Newsletter of the Australian Platypus Conservancy, Whittlesea, Victoria.
134. Anonymous (2005). Platypus go with the flow. *Ripples* Issue **30**, 2. Newsletter of the Australian Platypus Conservancy, Whittlesea, Victoria.
135. Anonymous (2005). Platypus move with the times. *Ripples* Issue **31**, 2. Newsletter of the Australian Platypus Conservancy, Whittlesea, Victoria.
136. Armit WE (1878). Notes on the presence of *Tachyglossus* and *Ornithorhynchus* in northern and north-eastern Queensland. *Journal of the Linnean Society of London* **14**, 411–413.

137. Barrett C (1944). *The Platypus*. Robertson and Mullens, Melbourne.
138. Bryant AG (1993). An evaluation of the habitat characteristics of pools used by platypuses (*Ornithorhynchus anatinus*) in the upper Macquarie River system. BSc (Appl. Sci. Hon.) Thesis. Charles Sturt University, Bathurst, Australia.
139. Grant TR (1992). Captures, movements and dispersal of platypuses, *Ornithorhynchus anatinus*, in the Shoalhaven River, New South Wales, with evaluation of capture and marking techniques. In *Platypus and Echidnas*. (Ed ML Augee) pp. 255–262. Royal Zoological Society of NSW, Sydney.
140. Grant TR and Carrick FN (1978). Some aspects of the ecology of the platypus, *Ornithorhynchus anatinus*, in the upper Shoalhaven River, New South Wales. *Australian Zoologist* **20**, 181–199.
141. Grant TR and Denny MJS (1991). Distribution of the platypus in Australia with guidelines for management. Report to Australian National Parks and Wildlife Service. Mount King Ecological Surveys, Oberon, New South Wales.
142. Grant TR, Grigg GC, Beard LA and Augee ML (1992). Movements and burrow use by platypuses, *Ornithorhynchus anatinus*, in the Thredbo River. In *Platypus and Echidnas*. (Ed ML Augee) pp. 263–267. Royal Zoological Society of NSW, Sydney.
143. Grigg G, Beard L, Grant T and Augee M (1992). Body temperature and diurnal activity patterns in the platypus (*Ornithorhynchus anatinus*) during winter. *Australian Journal of Zoology* **40**, 135–142.
144. Macquarie L (1810–1822). *Lachlan Macquarie: Governor of New South Wales. Journals of his tours in New South Wales and Van Diemen's Land 1810–1822*. Facsimile Edition 1956. Trustees of the Public Library of New South Wales, Sydney.
145. Rohweder D (1992). Management of platypus in the Richmond River catchment, northern New South Wales. BSc (Appl. Sci. Hon.) Thesis. University of New England Northern Rivers, Lismore, Australia.
146. Serena M (1996). Metropolitan monotremes. *Nature Australia* **25**, 28–32.
147. Serena M and Williams GA (1997). Population attributes of platypus (*Ornithorhynchus anatinus*) in Flinders Chase National Park, Kangaroo Island. *South Australian Naturalist* **72**, 28–34.
148. Waite ER (1896). The range of the platypus. *Proceedings of the Linnean Society of New South Wales* **21**, 500–502.

Chapter 7. Ancestry and evolution

[Also see references 2, 90].

149. Archer M (1995). Prehistoric platypus fits the bill. *Australian Geographic* **38**, 87–103.
150. Archer M, Hand SJ and Godthelp H (1991). *Riversleigh: The Story of Animals in Ancient Rainforests of Inland Australia*. Reprinted edition with revisions 1996. Reed Books, Kew, Victoria.
151. Archer M, Flannery TF, Ritchie A and Molnar RE (1985). First Mesozoic mammal from Australia – an early Cretaceous monotreme. *Nature* **318**, 363–366.
152. Flannery TF, Archer M, Rich TH and Jones R (1995). A new family of monotremes from the Cretaceous of Australia. *Nature* **377**, 418–420.

153. Long J, Archer M, Flannery T and Hand S (2002). *Prehistoric Mammals of Australia and New Guinea: One Hundred Million Years of Evolution*. University of NSW Press, Sydney.
154. Musser AM (1998). Evolution, biogeography and palaeontology of the Ornithorhynchidae. *Australian Mammalogy* **20**, 147–162.
155. Musser AM (2003). Review of the monotreme fossil record and comparison of palaeontological and molecular data. *Comparative Biochemistry and Physiology* **136A**, 927–942.
156. Musser AM (2006). Furry egg-layers: Monotreme relationships and radiations. In *Evolution and Biogeography of Australian Vertebrates*. (Eds JR Merrick, M Archer, GM Hickey and MSY Lee) pp. 523–550. Australian Scientific Publishing, Oatlands, New South Wales.
157. Musser AM and Archer M (1998). New information about the skull and dentary of the Miocene platypus, *Obdurodon dicksoni*, and a discussion of ornithorhynchid relationships. *Philosophical Transactions of the Royal Society of London Series B* **353**, 1063–1079.
158. Pascual R, Archer M, Jaureguizar EO, Prado JL, Godthelp H and Hand S (1992a). First discovery of monotremes in South America. *Nature* **356**, 704–705.
159. Pascual R, Archer M, Jaureguizar EO, Prado JL, Godthelp H and Hand S (1992b). The first non-Australian monotreme: An early Paleocene South American platypus (Monotremata, Ornithorhynchidae). In *Platypus and Echidnas*. (Ed ML Augee) pp. 1–14. Royal Zoological Society of NSW, Sydney.
160. Woodburne MO and Tedford RH (1975). The first Tertiary monotreme from Australia. *American Museum Novitates* Number **2588**, 1–11.

Chapter 8. Conservation: platypuses and people

[Also see references 6, 8, 9, 10, 18, 69, 128, 141, 144].
161. Anonymous (1954). The platypus in New South Wales. New South Wales. Fauna Protection Panel, Sydney.
162. Caughley G and Gunn A (1996). *Conservation Biology in Theory and Practice.* Blackwell, Cambridge, Massachusetts.
163. Collins D (1802). *An Account of the English Colony in New South Wales*. T. Cadell Jr and W. Davies, in the Strand, London.
164. Grant TR (1991). The biology and management of the platypus (*Ornithorhynchus anatinus*) in New South Wales. Species Management Report #5. NSW National Parks and Wildlife Service, Hurstville, New South Wales.
165. Grant TR (1993). The past and present freshwater fishery in New South Wales and the distribution and status of the platypus, *Ornithorhynchus anatinus*. *Australian Zoologist* **29**, 105–113.
166. Grant TR, Gehrke PC, Harris JH and Hartley S (2000). Distribution of the platypus (*Ornithorhynchus anatinus*) in New South Wales: Results of the 1994–96 New South Wales Rivers Survey. *Australian Mammalogy* **21**, 177–184.
167. Grant TR, Lowry MB, Pease B, Walford TR and Graham K (2004). Reducing by-catch of platypuses (*Ornithorhynchus anatinus*) in commercial and recreational fishing gear in New South Wales. *Proceedings of the Linnean Society of NSW* **125**, 259–272.
168. Harrison L (1922). Historical notes on the platypus. *Australian Zoologist* **2**, 134–142.

169. Home E (1802). A description of the anatomy of *Ornithorhynchus paradoxus*. *Philosophical Transactions of the Royal Society of London*, pp. 67–83.
170. Koch N, Munks SA, Utesch M, Davies PE and McIntosh PD (2006). The platypus *Ornithorhynchus anatinus* in headwater streams, and effects of pre-Code forest clearfelling, in the South Esk River catchment, Tasmania, Australia. *Australian Zoologist* **33**, 458–473.
171. Munday BL, Whittington RJ and Stewart NJ (1998). Disease conditions and subclinical infections of the platypus (*Ornithorhynchus anatinus*). *Philosophical Transactions of the Royal Society of London Series B* **353**, 1093–1099.
172. Munday BL, Stewart NJ and Sodergren A (2002). Accumulation of persistent organic pollutants in Tasmanian platypus (*Ornithorhynchus anatinus*). *Environmental Pollution* **120**, 233–237.
173. Otley HM and le Mar K (1998). Observations on avoidance of culverts by platypus. *Tasmanian Naturalist* **120**, 48–50.
174. Otley HM (2001). The use of community-based survey to determine the distribution of the platypus, *Ornithorhynchus anatinus*, in the Huon River catchment, southern Tasmania. *Australian Zoologist* **31**, 632–641.
175. Rosen S (1995). *Losing Ground: An Environmental History of the Hawkesbury-Nepean Catchment*. Hale and Ironmonger, Sydney.
176. Rutherfurd I and Abernethy B (1999). Controlling stream erosion. In *Riparian Land Management Technical Guidelines*. Vol 2. (Eds S Lovett and P Price) pp. 33–48. Land and Water Resources Research and Development Corporation, Canberra.
177. Serena M and Williams GA (1998). Rubber and plastic rubbish: a summary of the hazard posed to platypus *Ornithorhynchus anatinus* in suburban habitats. *Victorian Naturalist* **115**, 47–49.
178. Serena M and Pettigrove V (2005). Relationship of sediment toxicants and water quality to the distribution of platypus populations in urban streams. *Journal of the North American Benthological Society* **24**, 679–689.
179. Schiller CB and Harris JH (2001). Native and alien fish. In *Rivers as Ecological Systems: The Murray-Darling Basin*. (Ed WJ Young) pp. 229–258. Murray-Darling Basin Commission, Canberra.
180. Stewart NJ and Munday BL (2004). Possible differences in pathogenicity between cane toad-, frog- and platypus-derived isolates of *Mucor amphibiorum*, and a platypus-derived isolate of *Mucor circinelloides*. *Medical Mycology* **43**, 127–132.
181. Ward GM (1966). Once in the suburbs. *Victorian Naturalist* **83**, 157–167.
182. Whittington RJ, Connolly JH, Obendorf DL, Emmins J, Grant TR and Handasyde KA (2002). Serological responses against the pathogenic dimorphic fungus *Mucor amphibiorum* in populations of platypus (*Ornithorhynchus anatinus*) with and without ulcerative mycotic dermatitis. *Veterinary Microbiology* **87**, 59–71.

INDEX

Page numbers in **bold** refer to illustrations or photographs

Aboriginal
  diet 109–110
  knowledge of platypus 13, 109–110
  names for platypus 109
abundance 82–86, 138–139
activity 3, 64, 74, 87–88, 138
age 44, **45**, **46**, 133
agriculture effects of **39**, 124, **125**, 126, **127**
ancestry 8, 99–101, **102**, 103–106, 140, 141
Australian Platypus Conservancy 84, 92, 96, 122, 126, 134

behaviour
  aggressive 16, 31, 48–49, 136
  nest building 19, 22
  reproductive 17, **18**, 19, **20**, 136
  social 31, 134
  stress 132, 143
bill
  characteristics **5**, 27–28, **28**, **29**, 55, **56**, **57**, 133
  use in feeding 5, 52, 58–61
  nostrils on **5**, **56**, **57**
  sensory 52, 55, **56**, **57**, 57–61
blood 25, 67
bradycardia 66 *see also* diving
brain 53, **54**, 55, **57**, 58–61
breeding
  age at first 24
  biology 13–31
  season **16**, 16–17, **18**, 133–136
burrowing 8, 19, 52
burrows
  conditions in 19, 78, **79**
  nesting 19, 22, **23**, 24, 26–28
  resting 78–79, **79**, 86–89, **90**
  use of **40**, 86–89, **90**

captivity
  behaviour 17, 19, 29
  mortality 132, 143
  breeding 24, 26, 29, 74, 134
capture 83–84, **84**, **85**, 85–86, 135, 139, 141–142
carp 4, 119
caves, use of 23
cheek pouch 68, **69**, **70**, 71, **72**, 73
chromosomes 20–21, 135
cloaca 14, **15**, 19
conservation 107–129, 141–142
conservation status 11, 128–129, 141
copulation 17, **18**, 19, **20**
counter-current heat exchange 76, **77**
courtship 17, 19, **20**
crayfish (yabby) 52, 60, 69, 112, 137, 142
crural system **43**, 43–50, 135, 136 *see also* spur, venom, venom gland

dams, weirs 114–117
diet 66, 68–69, **70**, 71, **72**, 73–74, 137
digestive system 68
discovery 1, 107–108, **108**, 133
disease 120, **121**, 122
distribution
  current **9**, 10–11, 82, 133
  historical **9**, 10, 128, 141
diving 3, **33**, 63, 66, **67**, 68
drought 27, 91, **95**, 96–97, 140

ear 3, **5**, **33**, 53–54, 136
echidna
  crural system 43

egg 21, 22, **34**
fossil species 104, 105
hibernation 79
milk 25
reproduction 14, 21
species 1–3
hatchling 22, **34**
egg(s) 3, 13–14, 17, **18**, 19, 20, 21–22
envenomation 41–43, 135, 136
environmental impact **28**, **38**, **39**, 65, 87, 110, 112–124, **125**, 126, **127**, 128–129
evolutionary relationships 53, 103–105, 140
eye **5**, **33**, **34**, 52, 53, 136

fat
seasonal changes 78
storage 8, 91, 143
feeding *see* foraging
feet 5, 8
floods 91–92, **93**, **94**, **95**, 95–96, 140
food
items 68–69, **70**, 71, **72**, 137
quantity 26–27, 73–74
seasonal changes 71, **72**, 87
foraging 29, 52, 64–65, **65**, 69, 71, 138
forestry, impact of 123–124
fossil species 9, 53, 54, 59, 99–101, **102**, 103–105, 140, 141
fur
buoyancy 64
characteristics 8, 75–76, **76**
insulation 22, **36**, 75–76, **76**, 111
hunting for 14, **36**, 63, 111

gestation 17, 21
growth 27–28, **28**

habitat
degradation **38**, **39**, 87, 122–124, **125**, 126, **127**, 128–29
general **36**, **38**, **39**, 81–82, 139
requirements 64–65, **65**, 81–82, 86–87, 139

haemoglobin 25, 67
handling 42, 142–143
hatching 17, **18**, 22, **34**
Healesville Sanctuary 23, 27, 29, 74, 84, 134
heart rate 66
hibernation 79–80, 138
home range 88–89, **90**, 91, 140
hormones
adrenocorticoid/glucocorticoids 19
oestrogen 20
oxytocin 25–26
progesterone 16
reproductive 16, 20
testosterone 16

incubation 19, **22**, 23
injury 122–123, **123**, 142–143
intestine *see* digestive system
introduction
Kangaroo Island **9**, 9–10
Western Australia 10, 133

juvenile 27–28, **28**, **29**, 29–31, 133–134, 135
juvenile dispersal 30–31, 91, 135

kidney **15**

lactation 17, **18**, 25–27, 28–30
legal protection 111, 141
length 5, **7**, 27–28
longevity *see* age

mammary glands 25–26
mating 17, **18**, **20**
mating system 31, 49–50
metabolic rate 64, 75
milk
composition 25, 30
general 25–26
mortality
captive 49, 132
fishing **37**, 112–113, 141–142
natural 30, 68, 80, 94, 96, 115, 139
netting **84**, **85**, 142

road-kills 115, **116**
rubbish 122–123, **123**
movements 64, 83–85, 87–89, **90**, 91, 96–97
Mucor fungus *see* disease

nestling 25–27, **28**, **34**, 44, **45**, **46**
nostrils **5**, 52, **56**, **57**, 136

olfaction (smelling) **54**, 54–55
osmoregulation 117
ovary 14, **15**, 20

parasites *see* disease
pectoral girdle *see* skeleton
pelvic girdle *see* skeleton
penis 14, **15**, 19
pesticides 122, 126
plague minnow (*Gambusia holbrooki*; mosquito fish) 71, 119
pollution 113–114, 122, 126
population *see* abundance
population estimates 83–86
predators/predation 68, 80, 96, 115, 139

radio tracking 66, 74–75, 83, **90**, 118
redfin perch 119
reproductive system 14–15, **15**
rete mirabile 76, 77

salinity 117
sensory receptors
electro (electric fields) 55, **56**, **57**, 59–61
light (visual) 53, 58
mechano (touch) 55, **56**, **57**, 57–58
olfactory (smell) 54–55
sound (hearing) 53–54, 58
taste 55
sex determination 20–21, 135
sex ratio 134
sexual dimorphism 5, 7, 27–28
size 5, 7, 27
skeleton 3, 5, **6**
skull **6**, **102**
sleep 137
sperm 14, 16
spur **5**, **6**, 8, **35**, 41–44, **43**, **45**, **46**, 48–50, 136
swimming 3–4, 64, 138

tail **5**, **6**, 8, 42, 91, 143
Taronga Zoo 23, 48, 133, 134
taxonomy 1–3, 99–101, 103, 104, 132, 133
teeth 8–9, 101, **102**
temperature regulation 64, 73, 74–76, **76**, **77**, 78–80, 138
testes 14, **15**, **16**
ticks **35**, 120
trout 71, 73, 87, 112, 119–120, 135

urban development 122–123, **123**

venom 43, 47, 48, 111, 135–136
venom gland 8, **16**, 31, **43**, 47
vision 53, 58, 136
vocalisation 136

walking 8, 80
Warrawong Sanctuary 24, **29**, 50, 83
water rat 3, 7, 8, 73, 139
weight 5, 7, 27–28
willows 117–118, **118**, 119